D0264876

Stroud College of Further Education
T21657

TOTAL COST ASSESSMENT FOR ENVIRONMENTAL ENGINEERS AND MANAGERS

TOTAL COST ASSESSMENT FOR ENVIRONMENTAL ENGINEERS AND MANAGERS

Mitchell L. Kennedy

John Wiley & Sons, Inc.

New York ■ Chichester ■ Brisbane ■ Toronto ■ Singapore ■ Weinheim

This text is printed on acid-free paper.

Library of Congress Cataloging-in-Publication Data

Kennedy, Mitchell L.
Total cost assessment for environmental engineers and managers/ Mitchell L. Kennedy.
p. cm.
Includes bibliographical references and index.
ISBN 0-471-19098-5 (alk. paper)
1. Production management—Environmental aspects—Costs.
2. Pollution prevention—Costs. I. Title.
TS155.K43 1997
658. 15'53—dc21 97-17072

Printed in the United States of America

10 9 8 7 6 5 4 3 2 1

This book is dedicated to
Lauren
for her spark, encouragement and patience,
without which the book would not exist.

CONTENTS

PREFACE

My goal in writing this book was to provide a practical , no-nonsense guide for understanding the most persuasive tool in business: the financial analysis. While this tool has been used within management circles for decades, it only recently has been adapted for analyzing environmental projects. To become widely accepted among environmental professionals, the financial analysis process must be quick to use, easy to understand, and deliver accurate results in terms familiar to corporate financial officers and accountants. I have found Total Cost Assessment to excel in these areas, and in the last eight years watched it gain favor among practicing environmental consultants, industrial facility managers, and government agencies.

Despite the growing popularity of Total Cost Assessment, there was a lack of detailed information about the nuts and bolts of performing the analysis. The most vital aspect of the process, gathering the required data, had not been given more than a page or two in any of the many journal articles and publications about TCA. As a person "in-the-trenches," doing the data gathering and number crunching, I found this to be most frustrating.

This book also provides me the opportunity to include the much needed discussion of quantifying environmental liabilities, and integrating TCA with management paradigms such as ISO 14000, TQM, and Activity Based Costing. All of these recent quality and accounting structures offer synergistic frameworks that support both TCA and pollution prevention. It is just common sense to make use of these existing frameworks and not re-invent the wheel when trying to prove the financial benefits of a process change or account for environmental costs.

This book presents material as if you were conducting the TCA and learning how to do it as it is performed. Each chapter contains examples or exercises to reinforce the concepts and steps of a TCA. The exercises herein are pulled from my experiences at over 200 industrial facilities and make use of actual financial data and projects from those industries. The case studies highlight the financial benefits of pollution prevention (P2) projects rather than treatment or disposal technologies because P2 can be more challenging to justify despite its proven track record of success.

Through reading the text and following the examples you understand how to concretely demonstrate what is often known instinctively: that preventing pollution makes environmental and economic sense.

Acknowledgments

The author would like to acknowledge the following individuals for their support and contributions to the content and structure of the book; Emilio Abatecola, Guilio Balestrino, Bob Bell, Bill Bilkovich, Ken Carlson, Isabelle Cohen, Jim DeWitt, Paul Feltree, Matthew Fraga, Ken Geiser, Terry Goldberg, Robert Grimm, Dave Hartley, Bob Hilbert, Herb Hohn, David Kennedy, Judith Kennedy, Neil Levine, Jonathan Libber, Kevin O'Donnell, Mike Pilla, Frank Rogers, Scott Rossiter, Guthrie Sayen, Fred Scognamillo, Rob Snyder, and Greg Vasil.

Mitchell, L. Kennedy
July, 1997

CHAPTER 1

WHAT IS TOTAL COST ASSESSMENT?

POLLUTION PREVENTION: THE PHYSICIAN'S CHOICE

The physician's old adage "An ounce of prevention is worth a pound of cure" applies to today's environmental management issues just as it applies to keeping one's body healthy. Preventing pollution creates a cascade effect that relieves companies from other burdens connected with creating, treating, and disposing of industrial wastes.

Documented results of this preventative approach abound. One study[1] showed an average net benefit of $3.49 for each dollar invested in pollution prevention, resulting in 1.6 million pounds of waste reduced per project.

Current methods of financial assessment fail to capture this "preventative advantage" of many environmental projects and quantify it in economic terms. Whether it is an investment for a new industrial process or the building of a hydroelectric power dam, the total costs associated with the development, installation, and operation of the system are too often placed into overhead or disregarded. Yet, these forgotten costs ultimately make or break most prevention-oriented projects.

[1]INFORM, *Environmental Dividends: Cutting More Chemical Wastes*, New York: INFORM, 1992, p. 31.

Alan Donnahoe, CEO of Media General, sees overhead not as a fixed and accepted item, but a challenge to improve the company:

> These are the expenses that must be covered in good times and bad, and they determine the break-even point of the company. Thus all good managers are expected to work hard to control, and if possible to reduce, all the elements that enter into overhead expense.[2]

This line of reasoning makes an excellent argument for both pollution prevention and a new method of cost accounting. Often, a *supposedly* fixed cost such as annual waste disposal can be reduced or eliminated through pollution prevention. The replacement of chlorinated solvents with a "no-clean" alternative is one oft-touted example. In this case, all costs connected with the solvent can be eliminated, not just those lumped into overhead.

Unfortunately, it is difficult to know when a pollution prevention project will yield the large savings and when another project might be more preferable. The main obstacle is the lack of inclusion of environmental costs in the decision-making process. The INFORM study of chemical manufacturers, which reported the $3.49 benefit per dollar invested, also states that companies with some type of environmental cost accounting system implemented three times as many pollution prevention projects as those companies not tracking environmental costs.[3]

DECISION MAKING AND ENVIRONMENTAL IMPACT

When people think of cost accounting, several images may come to mind, including that of a bespectacled accountant behind mountains of receipts busily punching the keys of a calculator. Almost never does the image of a cleaner environment and safer workplace come to mind.

Why is it that no image of clear skies and healthy workers rushes to the forefront? Perhaps because traditional cost accounting does not measure environmental impacts, nor does it deliver cost data in enough detail to make sound decisions about environmental projects. For example, it has been

[2]Alan S. Donnahoe, *What Every Manager Should Know About Financial Analysis*. New York: Simon & Schuster, 1989, p. 42.

[3]INFORM, *Environmental Dividends: Cutting More Chemical Wastes*. New York: INFORM, 1992, p. 31.

shown on a macroeconomic scale that the Gross Domestic Product (GDP) of the United States can be increased by building more nuclear weapons, clear-cutting forests, or burning more fossil fuels. As a society, we know that jobs, paper products, and heating our homes are a necessary part of modern life, yet the impacts of these activities (nuclear waste and war, loss of wild habitat and species, and global warming) are not registered in the national economic system. The lack of a macroeconomic system to account for environmental damage simply reflects the lack of a microeconomic model for tracking the same issues. On a facility or plant level, the environmental impacts of financial decisions are rarely included in the project analysis. The following is a list of commonly used categories for cost accounting with which businesses conduct analysis and make capital purchasing decisions everyday:

COMMON ACCOUNTING CATEGORIES

Overhead	Direct labor
Inventory	Direct materials
Customer receivables	Equipment
Buildings	Bank debt
Maintenance	Payroll
Advertising	Utilities
Insurance	Taxes owed

Where in this list would the cost of keeping the company in compliance with all applicable environmental regulations be found? Or, how does the cost of disposing of hazardous wastes fit within these categories? If a corporation does not keep accounts of these costs, how easy will it be to argue the financial merits of a project that can eliminate hazardous wastes or cut environmental compliance tasks in half?

Most people, executives included, make decisions based on the information available or presented to them. If certain facts or costs are not included in the available information, these same facts never become part of the decision-making process. Facts are included when people who gather the facts have enough knowledge and experience with a new situation to know which facts are needed. Public concern for the environment and liabilities connected to polluting make up a relatively new situation compared with the length of time industrialization has been around. This newness can explain part of why such an "information gap" exists in the decision-making processes surrounding environmental projects.

Grouping environmental projects with other projects competing for the same limited capital expands this information gap because of the fundamentally different nature of environmental projects. No other capital purchase has a similar scope of repercussions, one that extends into the neighboring communities and local ecosystems, affects customers and suppliers, and has the potential to draw the positive or negative attention of federal, state, and local regulators, the media, and shareholders.

THE CALL FOR TOTAL COST ASSESSMENT

Clearly, a new financial analysis method that can more fully track and allocate the forgotten costs will allow companies to reap the benefits of pollution prevention and show that being green makes dollars and sense. There are myriads of decision-making tools, cost analysis software, and consultants willing to help you conduct financial analysis and capital budgeting. What you probably need is something simple, inexpensive, easy to use yet reliable, and having credibility with accountants and financial managers. This tool would account for the *total* operating costs of a process, environmental and otherwise, and would use common terminology and formulas found in the accounting profession. Total Cost Assessment (TCA) exceeds these requirements, yielding two additional benefits:

1. TCA enables managers and engineers to make more informed decisions about environmental projects. By having more information on chemical purchases, waste disposal, and other costs, a decision maker can see the benefits of pollution prevention. For example, if a decision were to be made about changing to a water-based printing ink, a traditional accounting of costs might show an increase in ink purchase costs, leading one to think the investment is not very prudent. Other printing operating costs would change also; some would decrease and potentially save the company more money than was originally invested. This would be undiscovered country using a traditional financial analysis tool.

2. TCA levels the playing field between competing capital purchases. An environmental professional trying to sell a project to management has little information to work with if the accounting systems have not segregated environmental costs. Other capital purchases often have more clearly defined costs that in turn, may make them look more attractive. For example, purchasing a delivery van can be clearly shown to have cost savings, including increased flexibility to ship finished products, pick up orders for processing, and decrease the likelihood of parts being damaged during transport. If the water-based printing ink is competing for scarce capital with a new delivery van, the benefits of the ink must be equal to or greater than the benefits of the delivery van. But without the detailed data that Total Cost Assessment provides, the numbers will not be visible to the decision makers. Conducting a TCA does not guarantee approval of the project, but it may give the project higher priority once the manager sees the costs involved or savings available.

HOW ARE TCA AND ENVIRONMENTAL COST ACCOUNTING RELATED?

Environmental Cost Accounting (ECA) relates directly to managerial accounting and internal reporting of the company's profitability. Managerial accounting is the branch of accounting that deals with financial decision making, investments, operations of the company, cost control, and budgeting. The other branch, financial accounting, develops the reports for external use by the shareholders and government. Although both branches do cost analysis, only managerial accounting is used to make decisions, and most companies will not share managerial accounting information with the public or government agencies.

Total Cost Assessment can be considered a tool used in the Environmental Cost Accounting arena. Both share many of the same cost categories and address the same goal of making costs of environmental impacts more visible to management. Table 1.1 shows some of the costs categories used in Environmental Cost Accounting.

Environmental Cost Accounting classifies these costs into conventional, hidden, contingent, and image costs. Most companies generally track conventional costs, and many track some of the hidden costs. Only about 30% consider just one of the contingent cost categories, with only 10% regularly attempting to quantify this third class of costs.[4] The fourth class, image costs, is rarely documented in conventional accounting systems. TCA attempts to bring all these cost categories (conventional, hidden, contingent, and image) to bear on whatever project is being analyzed in order to show all the benefits

[4]U.S. Environmental Protection Agency, *Environmental Cost Accounting for Capital Budgeting: A Benchmark Survey of Management Accountants*, EPA742-R-95-005. Washington, DC: U.S. EPA Office of Pollution Prevention and Toxics, 1995, p. ES-4.

■ TABLE 1.1 Cost Categories Used in Environmental Cost Accounting

Regulatory	Upfront	Voluntary
	Potentially Hidden Environmental Costs	
Notification	Site studies	Public relations
Reporting	Site preparation	Training
Monitoring /testing	Permitting	Audits
Studies / modeling	R&D	Qualifying suppliers
Remediation	Engineering & procurement	Planning
Record keeping	Installation	Feasibility studies
Plans	**Conventional Costs**	Remediation
Training	Capital equipment	Recycling
Inspections	Materials	Habitat & wetland protection
Manifesting	Labor	Landscaping
Labeling	Supplies	Financial support to environmental group
Protective equipment	Utilities	
Medical surveillance	Structures	
Financial assurance	Salvage value	
Taxes / fees	**Back End**	
Spill response	Closure/decommission	
Waste management	Disposal of inventory	
	Postclosure care	
	Contingent Costs	
Future compliance	Remediation	Legal expenses
Penalties & fines	Property damage	Natural resource damage
Response to future releases	Personal injury	Economic loss damage

(Continued)

TABLE 1.1 *Continued*

Regulatory	Upfront	Voluntary
	Image & Relationship Costs	
Corporate image	Relationship with professional staff	Relationship with lenders
Relationship with customers	Relationship with workers	Relationship with communities
Relationship with investors	Relationship with suppliers	Relationship with regulators
Relationship with insurers		

Source: U.S. Environmental Protection Agency.

or costs. A company having an Environmental Cost Accounting program will be able to easily extract cost data for the performance of a Total Cost Assessment. In this way, the two work synergistically. It is not necessary to have an Environmental Cost Accounting system in place in order to perform TCAs; it will just make the task of gathering the required data that much easier.

WHERE IS TCA TYPICALLY USED?

Total Cost Assessment can be applied to any project, environmental or otherwise, to yield comparative cash flows and projections of rates of return. Typical applications fall into four broad categories.

1. *Project Selection among Numerous Alternatives.* TCA can be used to weed out undesirable projects as a preliminary cut. This can be done using a shortened TCA method to avoid the need to gather large amounts of data on a large number of options.
2. *Project Final Analysis.* Using TCA to select one option out of a list of finalists or deciding if a single option is better than continuing to operate a process in the same manner is one of the more common applications.
3. *Supplemental Environmental Projects.* The U.S. Environmental Protection Agency (EPA) allows companies facing large penalties to propose mitigation projects that can offset a portion of the penalty. These projects must have a developed budget and timeline and the financial benefits to the company must be calculated using after-tax

net present value formulas. TCA is an acceptable method for quantifying these benefits.

4. *Budgeting for Alternatives.* Although this application is slightly more unorthodox, it nevertheless can provide first-cut figures on the maximum amount a company can spend on either the operating costs or capital purchases of a new system if the existing costs are not to be exceeded.

The need for TCA usually arises after the "low-hanging fruit" have been-picked. That is, some environmental projects are clear cost reducers and requre no calculations. Actions such as turning off a running hose or regulating the flow of rinse waters can save money the instant they are performed and do not need any justification. In fact, conducting TCA on these types of projects will probably take longer than doing them!

THE PURPOSE OF THIS BOOK

This book was written to educate environmental managers and engineers from the ground up on the benefits, applications, and mechanics of Total Cost Assessment. The goal is to provide enough information so that a person with little or no knowledge of accounting can analyze pollution prevention or other environmental projects . These tools will add credibility to the analyst's report and reveal the true costs and savings resulting from pollution prevention projects.

STRUCTURE OF THE BOOK

The book's structure allows the new reader learn about TCA as if performing one for the first time. Because of this, many examples are interspersed with definitions, and the section on gathering cost data precedes the section on analysis and calculation (just as it would if you were doing TCA). The book focuses on practical low-cost, low-time-demand methods for estimating and evaluating projects. These methods appeal to the broadest possible audience and provide a jumping-off point for those who desire greater detail. The exercises in the book are designed to teach skills in gathering data, performing the analysis, and evaluating intangible factors such as long-term liability. Later chapters show how TCA can be integrated with some of today's popular environmental management techniques such as ISO 14000, Theory of Constraints and Life Cycle Analysis. The book closes with a discussion of how to expand the methods of Total Cost Assessment beyond the walls of the facility and the macroeconomic construct.

CHAPTER 2

GATHERING COST DATA

This chapter discusses the singularly most important part of conducting a TCA: cost data. The computer programmers' adage "Garbage In–Garbage Out" applies here as it does in computer programming. The formulas used in the later chapters of this book cannot compensate for imprecise or missing data. In fact, the whole profitability of the project rests on the accurate and thorough gathering of cost data.

DEFINING DATA NEEDS

TCA requires data on operating costs for the existing system as well as the expected costs from the purchase and operation of the new system. As environmental manager, you must define the minimum number of data elements required to conduct the analysis and then locate or derive these figures.

Clearly, defining the required data elements provides the following benefits at the outset of the TCA:

1. By defining the data required, the analysis becomes more focused, revealing potential problems or considerations. A well-focused analysis will require less data and less effort. The thought of performing a TCA might intimidate some people who may fear that the analysis will take too much time and energy. Because of this notion, people may be resistant to providing information or assisting in the analysis. Clearly defining the parameters to be analyzed helps dispel this notion and reassure people that this is not such a time-consuming exercise.

2. Defining the data required also saves time when collecting data. For example, by reviewing historical process data, it may be possible to narrow the search from the previous 5 years to just the past year if you know what is needed ahead of time. This saves time twice: once when collecting data (especially if the records are achieved off-site or on computer tapes) and then during analysis when fewer numbers require processing.

This first step also sets limits for the level of expected detail. Significant time can be spent locating or deriving minutia that have insignificant impacts on the end result. If this level of detail is required by upper management or corporate policy, then the time is warranted; otherwise follow the 80–20 rule. This is also known as the law of diminishing returns. This axiom states that 80% of the result requires 20% of the effort and the remaining 20% requires the additional 80% of the effort. Although this may seem to run counter to the philosophy of extracting total costs, remember TCA is a practical tool and not an academic exercise. As you perform the TCA remember to maintain the balance between insufficient detail and overkill. Lack of detail can decrease the accuracy and credibility of the analysis. Obsessive detail can overtax valuable labor hours.

Once the scope and type of data have been developed, the data must be uncovered and collected. Gathering cost data requires creativity and persistence. Companies by and large do not separate costs to the level of detail required by a TCA. Unless the company has adopted Activity Based Costing (see Chapter 10), costs tend to be allocated across broad categories such as marketing, facility maintenance, labor, materials, inventory, and overhead. Whether a company follows Generally Accepted Accounting Principles (GAAP) does not matter very much. Most companies follow GAAP for financial accounting, but TCA requires data from managerial accounting systems, which may or may not shadow the GAAP norm. If managerial accounting systems do follow GAAP, chances are good the data required still will not be in the format needed for the TCA.

Managerial cost accounting practices can vary tremendously from company to company based on corporate directives and personal preference for methods and categories. Keep in mind these systems have evolved to deliver the data demanded by management while minimizing the effort required to generate this data. Therefore, if the data element you seek does not appear as a cost category, it will have to be derived from the data that are already collected or measured directly at the manufacturing process.

For example, most companies retain their utility bills, so it should be relatively simple to find out the amount of electricity used by the facility in 1996 and the associated cost. The question "What is the electricity usage of the green vapor degreaser in Department 007?" is less likely to have a ready answer, not because the vapor degreaser did not use electricity, just that no one in the company had the need to know that particular number. In this case, knowing the limits of the accounting system you are working with will save time and avoid the frustration of asking for something that does not exist.

THE USEFULNESS OF MATERIALS TRACKING SYSTEMS

A materials tracking system is a mechanism for determining where and how much of a chemical is used within a facility, and can be a centralized location to store information about materials, such as vendor names, unit prices, delivery dates, and associated environmental risks and hazards. Over the past 20 years, two major factors have made materials tracking systems a necessary part of being a viable and competitive industrial plant.

1. *Growth of Environmental and Worker Safety Regulations.* Through the U.S. Environmental Protection Agency (EPA) and the Occupational Safety and Health Administration (OSHA), the federal government has forced industries to keep records of the materials that workers are exposed to and/or emitted into the environment. Annual and semiannual reports such as Superfund Amendment and Reauthorizatioin Act (SARA) 311, 312, 313 and the Resource Conservation and Recovery Act (RCRA) Biennial Hazardous Waste report require companies to total the quantities of chemicals on-site at any one time, the maximum amounts of materials stored on-site, and the amounts of chemical by-products released to the land, water, and air. To comply with these regulations, companies have had to begin tracking their use of materials through their facilities. At the same time, the government has required chemical manufacturers to distribute Material Safety Data Sheets (MSDS) on each hazardous chemical sent to a user. Typically, this MSDS contains four pages of information on physical properties, health hazards, storage and use precautions, and emergency measures. It was not long before facility managers had

2- and 3-foot piles of these MSDSs taking up shelf space. Software developers recognized an opportunity here and soon developed software containing MSDSs for all hazardous materials, as well as report writing systems for SARA and RCRA reports.
2. *Quality Management Paradigms.* A growing movement toward more precise process control and material tracking forced change from the other direction. This meant a greater ability to replicate ideal design specifications or customer standards. Dr. Walter Demming first introduced the concepts of statistical process control in the late 1940s. At the time, it was not well received in the United States, so he taught his methods to the Japanese, who readily adopted them. From Demming's models grew other parallel practices such as Just-in-Time (JIT) manufacturing, Total Quality Management (TQM), and most recently the ISO 9000 movement.

Just-in-Time manufacturing strives to eliminate inventory by manufacturing only what the customer orders. Parts for a desk, for example, would be created and assembled at the same time, rather than having a stockpile of legs and drawers or other subassemblies. This philosophy demands accuracy in materials tracking to ensure that all parts required for the unit are flowing through production at the right time.

Total Quality Management promotes continuous improvement of all manufacturing operations through a team review of specific production processes or product lines. This periodic and regular review promotes incremental change focused mainly on meeting production goals, quality standards, and reducing reject rates. This process of ongoing review forces examination of materials usage, work flow through the facility, trouble areas, characteristics of the materials used, and worker habits. Greater control over the flow of materials, quality, and number of materials is a natural outgrowth of this process.

The International Standards Organization (ISO) recently promulgated unified quality codes for international manufacturing operations. These are referred to by their code number, ISO 9000. All manufacturing corporations desiring to sell goods in the European Union are required to have ISO certification for these codes. The theory here is that the codes will improve the quality of goods manufactured by forcing a minimum standard and will level the economic playing field by excluding those who would sell low-quality goods at lower prices. Receiving ISO certification requires a rigorous external review of accounting, purchasing, shipping and receiving, processing, and quality inspection systems. Each system has a list of checks and balances that must be implemented to receive the certification. All of these changes must be documented and periodically reviewed to maintain certification. As a result, ISO-certified companies end up installing tracking systems or portions of a full system to achieve and maintain the certification.

In response to environmental regulations and product quality issues, many industries have developed or purchased materials tracking systems. There are as many varieties of materials tracking systems as there are accounting systems. These range from home-grown databases built in Paradox or Excel software to one of the over 120 commercially available software packages such as AV Systems' Material Inventory Report System and Quantum Compliance Systems' Facts. If the facility you are working with has such a system it is advisable to spend several hours learning how it tracks the materials; finding its smallest unit of measure for tracking; determining if it also includes costs, shelf life, delivery dates and quantities, and vendor information.

The best systems will have all this information broken down by product line or production process code. All these factors affect usage and cost and can highlight problems with existing production processes. Although no TCA will be performed as part of the following example, it highlights the usefulness of reviewing materials inventories and potential cost savings.

A large metalworking company reviewed its water-soluble machine tool coolant usage and found that over two dozen varieties were used. The company had to keep all of these in stock (which occupied significant storage space), maintain each separately to keep them from fouling (which involved considerable labor for testing and adding chemicals), and properly use each one as specified (which required separate instructions and trained personnel). A comparison of each coolants' lubricity packages and additives resulted in an opportunity to reduce the coolant selection from two dozen to under 10. Cost savings attributable to reviewing materials use were estimated to be $20,000 to $40,000 annually.

LOCATING COST INFORMATION

The types of cost information used in a TCA are usually of two types: costs connected to the purchase of new equipment and the day-to-day operation of the process, and ancillary costs that are impacted by the process and chemical use, but are not a part of production. Examples of the former are chemical purchases and electrical energy use for a particular production process. Examples of the latter are the impact the chemical has on insurance premiums and the additional waste treatment needed.

■ TABLE 2.1 Environmental Cost Accounting Tiers and the Cost Categories of TCA

Conventional Costs	Hidden Costs	Contingent Costs	Image Costs
Chemical purchases	Ancillary chemicals	Covered in Chapter 9	Covered in Chapter 9
Chemical storage	Waste management		
Process maintenance	■ Testing		
Utilities	■ Treatment		
	■ Disposal		
	Regulatory compliance		
	Fees & taxes		
	Insurance		
	Production		

This book follows the classification developed by the EPA for Environmental Cost Accounting. These "tiers" are labeled "Conventional Costs," "Hidden Costs," "Contingent Costs," and Image Costs," and contain smaller cost categories within them. The following description of the first two cost categories presents each one according to where it falls in the Environmental Cost Accounting tiers (see Table 2.1). Costs within the "Contingent" and "Image" tiers are discussed later.

Total Cost Assessment creates a fine-grained level of detail that allows costs of production to be attributed to individual chemical use or production processes. Because of this, the cost categories are often process- or chemical-specific, instead of department- or function-specific. For example, the exercises in this book refer to labor connected to the maintenance of a specific process, not the maintenance of the facility, and the disposal costs refer to the disposal of wastes for a particular product or process, not the entire waste stream of the facility.

COST CATEGORIES

The cost categories typically used in a TCA are detailed in what follows. Most of these categories are used in the case studies. Others like ancillary chemicals, storage space, and insurance costs may appear only on the worksheets for which these categories have values. The reader should explore the applicability of each cost category to a specific situation and develop a customized list. Each category description that follows includes who within the company may have the data needed and which records to examine when gathering cost data. The EPA tier for each category is provided in parentheses after the name of the cost category.

Chemical Purchase Costs (Conventional)

The first and most obvious category is the use of chemicals and materials. Develop a list of all chemical inputs required by the process. This can be facilitated by a process flow diagram, engineering drawings, and a walk-through or -around the process in question. Research the usage and cost of each input. The purchasing department typically maintains files on chemical and materials purchases. Medium to large companies will undoubtedly have computerized material tracking or inventory systems in place. For smaller companies, or those without computerized tracking systems, collecting chemical usage data requires more legwork. Talk first to the production supervisor in charge of the area where the chemicals and materials are used. Find out information such as trade names and departmental requisition numbers that would be used to track the materials in question. This will enable you to effectively communicate with the purchasing department, minimizing wasted time and frustration.

Next, visit the purchasing department and present a list of the desired information. By having a written list, the purchasing manager does not have to remember everything asked for, and you can be sure to receive all the information requested. Be prepared to explain why you want this information in nonstandard formats, such as unit prices or allocations to specific departments. A company with a vibrant pollution prevention team will already have a member from purchasing involved, making this step in the TCA more efficient.

What can you do if purchasing cannot help? Departments other than purchasing may have copies of internal requisition slips or purchase orders that may have price and quantity data. Vendors who supply the company with the product will generally furnish current market prices if no other record can be found. In some cases, these vendors also may be able to summarize their customer's accounts and offer a printout of total quantities of materials purchased to date. Be aware that the purchasing department may have changed suppliers midyear, and you may need to consult more than one vendor for the complete picture. Differences in formulation concentration and additives may also effect the cost of materials. Make note of these differences as they could be relevant in the overall discussion of options. For example, some companies have switched their cleaning solvent from 100% mineral spirits to a blend of surfactants with a small percentage of mineral spirits. If the main focus of the project is to stop using all mineral spirit products, the data gatherer would want to note that the selected alternative is not 100% free of mineral spirits. This could become an important consideration when comparing other alternatives.

Compile all materials usage figures into annual totals and develop unit prices. This extra step allows direct comparison between alternatives and can show how much of each material has been or will be used per year. This issue may be a concern if storage space for the material is limited or if threshold limits such as SARA 311, 312 and 313 reporting (see Glossary) could be exceeded. Determining annual usage helps check for chemicals that should be reported under the aforementioned laws but for whatever reason are not. The EPA has established daily fines ($25,000/day per chemical) for not reporting listed chemicals when usage exceeds regulatory thresholds. These penalties provide strong incentive to check all chemical usage levels.

Chemical Storage Costs (Conventional)

Many facilities rent warehouse space to store raw materials, spare parts inventory, and finished products. In these cases, a change in production processes that reduces the need for raw materials or changes the form in which the raw material is used (e.g., using a gas instead of a liquid or solid) may result in decreased need for storage space. In other cases, the warehouses are owned by the company and are carried as a financial asset whether full or empty. Floor space within the production facility is usually at a premium, creating conditions for cost savings if additional warehouse space becomes available.

For example, an electroplater could have added a fourth plating line and increased sales if the chemical storage area had not taken up the extra space. Because the company had already built out to the property lines, and vertical expansion was not feasible, this storage space acquired a high value. When an opportunity to change production and decrease chemical usage also reduced storage space requirements, the company's environmental engineer included this fact in the analysis. The potential revenue from a proposed fourth plating line was not a direct line item but a footnote with an estimated, albeit eye-catching value. In these instances, the actual amount of revenue would be difficult to estimate, but the TCA should include such a footnote about the opportunity to expand production. Ask the plant manager or shipping and receiving department to identify which warehouses the materials in question are located. Knowing the location, the manager/engineer can ask the company accountant how much the company pays for this space. Visit the warehouse and measure the space occupied by the material, or estimate the square feet occupied based the dimensions of the materials' containers.

$$\text{Storage costs} = \text{square feet of storage occupied by the material} \times \text{unit cost } (\$/\text{ft}^2) \text{ for warehouse space}$$

The case studies in this book include storage costs whenever there were actual costs involved. Include this cost category to stimulate discussion of alternative uses for the storage space. This discussion may generate additional ideas for pollution prevention or reductions in inventory. Perhaps one alternative will not result in enough space savings to accommodate additional production equipment, but the cumulative space savings of several ideas derived from the discussion may. Do not get sidetracked from analyzing the original project. Footnote these other possibilities in the TCA summary.

Be aware that first looks into a production process often result in many potential pollution prevention projects. Excessive warehouse and storage costs are a red flag to the environmental manager that a materials usage opportunity may exist. Exercise caution about which process requires its materials or products to be stored, and whether allocation of this space to the current TCA is relevant. In other words, save these material storage opportunities for another day if the current project does not impact it. It would be inappropriate, for example, to include the storage cost savings from the following three options if only the first one was the subject of the TCA.

- Instituting a program for regular review of material shelf life and inventories
- Installing vertical storage bins
- Having the materials supplier delivery half as much material twice as often.

The manager/engineer must be consistent in the allocation of costs. There is no reason why all three of the options could not be included in a TCA as long as all the other costs connected to each option were also included. The case studies in this book either carefully define storage costs for only the chemicals required by the process under analysis or omit this cost altogether if not applicable or unclear. Increases in production capacity due to the availability of more floor space are only included if the project will result in enough additional floor space to add a complete additional process or operation. Note that if the analysis includes cost savings from the gain of floor space, it must also include the costs for purchasing the additional production equipment. This may be beyond the scope of the original TCA and requires careful consideration by the manager/engineer on a case-by-case basis.

Process Maintenance Costs (Conventional)

In this book, the term *maintenance* refers to maintenance performed on the process being analyzed. This maintenance may include lubrication, replacement of pumps, fittings, filters, nozzles, gaskets or other replaceable parts; and painting, cleaning, emptying, refilling, daily additions of chemicals, or testing of bath concentrations. This maintenance may take place hourly, daily, monthly, or annually. It is process-specific and deals with maintaining the proper functioning of the machinery and equipment used in the process. It does not include general grounds keeping, floor sweeping, or general rubbish removal unless these activities can be directly related to the process being analyzed. A pollution prevention project that decreases the rubbish accumulation plantwide would need to account for general plant maintenance, as would any process that resulted in less deterioration of building components. A process change that eliminates corrosive fumes might decrease the corrosion to structural building components within the facility or in the area adjacent to exhaust vents.

The best source for information on maintenance costs is the maintenance foreman, shift manager, or process operator responsible for keeping the process running. A brief discussion of responsibilities regarding maintenance should yield the type of activities performed, when they are performed, why they need to be done as often as stated, and how long each task takes. Note the cost of any replacements parts, how many, and how frequently. If the maintenance foreman cannot furnish the information, purchasing or accounting should have records of maintenance materials requested. The personnel department can furnish labor rates. Determine the total maintenance cost of the existing system by compiling all material costs (annualized) and the total labor costs (total of labor for each task multiplied by the labor rate for each employee performing the task). Generalized labor rates may be used to reduce the number of separate calculations but should be footnoted in the TCA report.

A Maintenance Cost Example Every 6 weeks, the water-soluble metalworking coolant in a CNC mill has to be changed because it is rancid (develops a foul smell from the proliferation of anaerobic bacteria). This 50-gal coolant sump must be drained, flushed with a machine tool cleaning agent, and refilled to the proper concentration with fresh coolant. These operations take one employee 2hr to perform. The average wage at the shop is $14.00/hr, coolant concentrate has an in-sump cost of $0.40/gal, and the machine tool cleaner is $1/gal. Do we have all the information necessary to calculate the total maintenance cost of this machine coolant? What is missing?

The amount of machine tool cleaner used was not stated. If this was given as 20 gal, then the maintenance cost every 6 weeks for the machine sump would be $40 in materials and $28 in labor.

$$\begin{aligned}\text{Labor cost} &= 2\text{ hr per change} \times \$14.00\text{ /hr per person} \times 1\text{ person} \\ &= \$28.00\text{ per changeover}\end{aligned}$$

$$\begin{aligned}\text{Material cost} &= 50\text{-gal sump} \times \$0.40/\text{gal coolant} \\ &\quad + 20\text{ gal} \times \$1/\text{gal cleaner} \\ &= \$40\text{ per changeover}\end{aligned}$$

Over a full year (8.66 coolant changeovers), the total maintenance cost is $590.

There are several subtleties to note in this example. The manager/engineer would have to know at what concentration the coolant is mixed. For a water-soluble oil, this is usually 20:1 or 5%. Therefore if the purchase cost of the coolant concentrate were given instead of an in-sump cost, it would be necessary to either know how much pure concentrate is used each time or compute a unit price for the coolant when mixed with water and placed in the machine sump. Note that the cost of the coolant is included as a maintenance cost rather than a chemical use cost. This would be acceptable as long as it was not recorded in both categories, resulting in double counting.

Questions should be asked about where the waste coolant and machine cleaner go. Are they flushed down a drain? Are they placed in drums for off-site disposal or recycling? These are additional costs to incorporate in the TCA. Most importantly, why is the coolant changed so frequently? It is quite possible that such frequent changovers are considered standard practice and have never been questioned. Coolants and management techniques exist that allow extended use of the same coolant for 1 year or more. The costs and benefits of these, of course, would be a subject for a TCA.

Utilities Costs (Conventional)

Quantifying the costs of utilities is the most difficult part of gathering cost data. Although larger companies tend to have better data-tracking systems, the larger size of the facility, and the number of buildings, machines, and processes offset this advantage. In fact, it is often easier to determine utility usage of a specific process within a small company because that process might be the major energy user.

In a small electroplating facility, the rectifiers for the plating baths might be the major consumer of electricity, even larger than facility lighting. The difficulty stems from the lack of individual metering at each process. From the water, gas, or electric company's viewpoint, all the water, gas, or electricity is flowing to one destination and what happens once it is inside the building is of no concern. From a corporate management or accounts payable department viewpoint, one utility meter is all that is needed to pay the monthly bills. Additional meters would simply complicate the process. However, from a pollution prevention viewpoint, it is vital to know how much water, gas, or electricity is being used by a process so that positive changes to that process can be evaluated.

Water Usage Costs

Water usage is typically billed with sewer usage, under the assumption that what flows into a facility must flow out of a facility. Exceptions exist such as evaporators, closed-loop systems, and so forth; see the following section on "Sewer Use Fees". For smaller facilities with a single water-using process, the quantity shown on the bill (usually in 100-ft^3 increments) is the quantity used and cost incurred. Sanitary water usage is relatively insignificant, but can be estimated by multiplying the number of toilets in the facility by an average flush amount (4 to 8 gal) and frequency (2 to 3 times per day per employee). To illustrate this claim, review the following example.

A Water Usage Cost Example A 20 person electroplating facility may have a sanitary usage of 160 to 480 gal per day. This is based on each person using the facilities two times per day with the previously stated flush amount of 4 to 8 gallons.

$$\text{Sanitary water use} = 20 \text{ people} \times 2 \text{ trips/day} \times 4 \text{ to } 8 \text{ gal per flush}$$
$$= 160 \text{ to } 480 \text{ gal/day}$$

The combined flow of the rinses from the two electroplating lines could range from 3,000 to 9,000 gallons per day for rinses running between 0.5 and 2 gal/min.

$$\text{Process water use} = 2 \text{ plating lines} \times 6 \text{ running rinses}$$
$$\times\ 0.5 \text{ to } 2 \text{ gal/min each} \times 60 \text{ min/hr} \times 8 \text{ hr/day}$$
$$= 3{,}000 \text{ to } 9{,}000 \text{ gal/day}$$

$$\text{Total water use} = 160 \text{ to } 480 \text{ gal/day} + 3{,}000 \text{ to } 9{,}000 \text{ gal/day}$$
$$= 3{,}160 \text{ to } 9{,}480 \text{ gal/day}$$

As a proportion of total water use, sanitary usage is at best 14% and as low as 2% for this small example. Note that as the plating lines become more efficient in using their rinsewaters (0.5 gal/min flow), the sanitary usage is a more significant part of the total water use. An increase in the concentration of contaminants is a typical problem when completing water conservation projects, and may even throw a company out of compliance with effluent limits.

More Complex Water Measurements

If the facility you are working with is more complex, several other estimating methods are available.

1. The bucket and stopwatch method is crude, but can give fast results with minimum effort. Simply time how long it takes to fill a measured or graduated bucket placed beneath a valve, inflow, or outflow pipe of the process in question. Assuming this flow rate is constant and the volume of the bucket is known, the flow rate can be determined by dividing bucket volume by elapsed time. A 5-gal bucket, filled in 15 sec would equate to a 20 gal/min flow rate.

$$\text{Flow rate} = 5\text{-gal bucket} \times 15 \text{ sec to fill} \times \text{four fillings/min}$$
$$= 20 \text{ gal/min}$$

2. Closing off the inflow, draining a holding tank, and then allowing it to refill will also provide a flow rate. The volume of the tank can be calculated from its inside measurements and using the stopwatch to record the time required to refill the tank. If a 3-ft by 5-ft by 3-ft deep tank requires 15 min to fill, then the flow rate is

$$\text{Flow rate} = \text{volume of tank } (45\ \text{ft}^3) \times \frac{7.48052\ \text{gal/ft}^3}{\text{time required to fill tank (15 min)}}$$
$$= 22\ \text{gal/min}$$ [1]

3. Manufacturers specifications on pumps and nozzles can also yield a flow rate if other variables such as leaks and evaporation are ignored. A 0.5-gal/min flow-restricting nozzle on a rinse tank provides the flow rate, and the total quantity of water used can be estimated by multiplying this rate with the number of hours the rinse tanks are running. If a restriction setting on the incoming water nozzle is stated as 0.5 gal/min, then the flow rate can be assumed to be the same as the setting provided the tank is not filling up or emptying.

At larger facilities, managers/engineers can use plant schematics, engineering departments, and portable flowmeters to determine mathematically flow and water usage for a specific process. Through the use of pipe diameters, pump horsepower, and gross water usage process water use can be determined. Engineering specifications on cooling towers and evaporators along with the hours they are operational can quantify water loss rates from these systems.

Natural Gas, Oil, and Coal Costs

Natural gas, oil, or coal usage is more easily determined than water because there is little evaporative and line loss. Process heating equipment and building space heaters almost always have total Btu or Btu/hr ratings stamped on their manufacturer's plates. This, in conjunction with standard conversion factors for heat content of fuels and the hours of operation, will provide the amount of fuel consumed. Equipment vendors also may be able to supply the fuel usage information if the equipment is still produced or they have serviced the same type of equipment since it was manufactured.

Storage costs may be a consideration for fuels such as oil and coal. Often times, these fuels require tanks or bunkers ranging from 5,000 to over 100,000 gal. Large fuel bunkers trigger compliance requirements for several federal environmental laws. These laws require continuous maintenance and monitoring of the storage devices, and spill-response training for employees. All these activities incur a cost that can be quantified (see the following sec-

[1] Conversion factors from *The Conversion Handbook*. Santa Fe, NM: Controls for Environmental Pollution, 1983.

tions on Safety Training and Equipment) and added to the TCA. The potential for a large spill that damages the surrounding water supply or ecosystem creates an additional liability to be included in the TCA. Liabilities from potential accidents are more thoroughly discussed in Chapter 11. If time or desire does not permit additional calculations, suffice to footnote the TCA that these liabilities exist and should be considered when reviewing alternatives. The severity of damage and repercussions of oil spills can justify the extra preventative measures and their associated costs.

Cost of Electricity

The cost of electricity used by a process is one of the most difficult costs to determine with a high degree of accuracy. As mentioned before, metering of utilities is done at the point they enter the facility. Some industries like electroplaters have process controls and gauges that show the ampere-hours consumed or watts being used, but most have no process-specific metering. There also tends to be more electrically driven process components than fuel-oil-, coal-, natural-gas-, or water-consuming components.

In textile dyeing, for example, electricity is required to turn each roller that unwinds or winds the fabric, electricity runs the blowers that exhaust fumes from the room, and electrical pumps mix dye solutions and deliver water to the dye jigs. Electricity also lights the room so the workers can see what they are dyeing. By comparison, water is used only in the first half of the process, when the fabric is being dyed, and gas or oil is used only in the second half of the process, when the fabric is being dried. To calculate the electricity costs for this dyeing process, operational information such as the size of the electrical motor, hours of operation, and type of electricity delivered to it (i.e., three-phase, 220 volts, or 110 volts, etc.) would have to be collected for each electrical device. Collecting this level of detail raises concerns about the 80/20 rule. Is it possible to identify all the electrical devices in the process and collect data on just the largest ones? In the textile dye example, the motors driving the unwinding and winding spools are the largest users. These motors are probably of the same type and model, allowing the manager/engineer to collect data on one and multiply by the total number of motors. This method also can be used for measuring water use in a long line of tanks (such as in electroplating or textile dyeing, or circuit-board fabricating). The largest electricity user has been determined. The other users, such as the lights, exhaust fans, and blowers, could be determined the same way, or not included all together if their contribution seems minimal. Best engineering judgment should be used here. Document all the assumptions in either case.

Another method involves the use of an amp-clamp, which is essentially a portable amp meter. This technology can read the amperage flowing through a line, and provide the data needed for total electricity use. Many industrial supply houses carry amp-clamps for under $600.

Large and/or Special Electrical Equipment Most specialized electrical equipment, such as vapor degreasers, baghouses, machine tools, infrared or ultraviolet drying/curing machines, and so forth, will have manufacturer's plates describing the appropriate rating of volts and amperes. Using basic formulas such as

$$\text{volts} \times \text{amperes} = \text{watts}$$

will yield electricity usage in watts at any given moment that the device is on. This rating can be multiplied by the number of hours that the equipment operates per day, month, or year to determine the "watt-hours" of usage. Divide this by 1000 to determine "kilowatt-hours" (kWh). The facility's electricity rate is based on kilowatt-hours, usually some number of cents per kWh (i.e., $0.10/kWh). Simply multiply the usage with the rate to determine the electricity cost for the specific device. Summation of these calculations will yield the total electricity cost.

Another less precise approach is to interview the foreman or plant manager and ask for estimates on the amount of time the process in question operates, what percentage of the facility's electrical demand results from the process, and if the process is run all year round or only during particular months. With this information and a copy of a utility bill from the accounting department, electricity use and cost can be approximated.

ESTIMATING ENERGY USE WITH CHEMICAL FORMULAS

Chemical formulations can be used to estimate energy required for drying. For example, a process simultaneously forms calcium chloride, water, and carbon dioxide. The carbon dioxide is released to the atmosphere, but energy is required to drive off the water. The theoretical amount of energy required can be calculated using the latent heat of vaporization for water, the ratio of moles of water to moles of calcium chloride produced by the reaction, and the total amount of water produced. Again, this number would be theoretical under ideal conditions. Many variables would impact the reaction and the drying of the product, most likely increasing the amount of heat energy required to dry the salt. Once the heat energy is calculated, it can be easily converted to Btus. Average drying time for a single batch can be used to determine energy input per unit time and the fuel required to deliver this rate of energy input. A unit cost for fuel multiplied by this quantity will yield a minimum energy cost for the drying process.

Usage Fluctuation

With all utilities, it is important to find out how many hours per day the electricity, oil, coal, gas, or water is on, and if the process shuts down during a particular time of the year. Specialty finishing processes, kilns, ovens, and other dryers are commonly shut down when not in use to conserve energy. Some food-processing companies are seasonal, running condensers, evaporators, heat exchangers, and other process equipment only when the fruit, grain, or meats arrive at their receiving dock. This may be only from September through March or from June through October depending on growing seasons and harvest times. Obviously, utility usage during downtimes is artificially low. A quarterly or monthly utility bill would not show the typical loads imposed when the equipment or process operates. If this were overlooked, the operating costs for the current system would be artificially low, resulting in reduced cost savings when compared to an alternative process. Besides detracting from the financial feasibility of the alternative process, the error could cost the credibility of the manager/engineer if discovered.

In the food-processing industry, a poor growing season can shorten production runs and give the appearance of low annual energy use. In these instances, average the utility usage across several growing seasons to provide a margin of safety in the cost figures. Also, ask the foreman or plant manager how many hours, days, and months on average does the process run.

Ancillary Chemicals (Hidden)

Do not overlook intermediate or ancillary chemicals and material inputs. These inputs may be affected by the change in a manufacturing process and either increase or decrease operating costs. Torberg considers these effects to be "cross-linkages" that can shift projects effectively from poor financial performance to superior.[2]

For example, a stamping operation uses a heavyweight petroleum lubricant during the stamping process. The company must clean the processed parts to remove this heavy stamping oil. If the company then changes to a lighter vanishing oil, this decreases the need to clean the stamped parts. Note here that both the purchase cost of the old and new stamping oils and the ancillary cost of the cleaning chemicals should be captured because the type of oil used affects the amount of cleaning required. There also may be waste-disposal costs associated with the ancillary chemicals. In the preceding example, the waste solvent from the parts cleaning operation could be a hazardous waste, and could increase in quantity and cost depending on the type of stamping oil being used.

As another example, consider a company that paints its products using a solvent-based paint. If the company chooses to change their painting processes from a solvent-based system to a water-based system, there would be obvious decreases in air emissions, and potential cost savings from the price of the water-based paint versus the solvent-based paint. Every painting operation requires cleaning of the painting equipment and the paint booth area. In this case, the ancillary solvents and cleaners used to clean the painting equipment also should be considered. Waste solvent and waste-solvent-based paint would be most likely considered a hazardous waste.

Storage, disposal, and tracking of a hazardous waste will always increase costs more than a nonhazardous waste. Many water-based systems can be cleaned up with soap and water, and the rinse water treated prior to discharge to the sewer. There will be some cost associated with treating this water-based paint rinse water (especially if the company has no other water discharges). These cost differences must be accounted for and entered into the TCA.

Waste Management (Hidden)

The second most significant cost category in TCA is impacts from waste-management activities. These costs include all treatment chemicals such as sodium hydroxide and sulfuric acid for pH adjustment, all tests and analyses performed to ensure effluent or wastes meet expected criteria, and all disposal costs connected with removing the waste from the facility.

[2] Richard Torberg, "Capital Budgeting for Environmental Professionals," *Pollution Prevention Review*, Autumn 1994, p. 462.

Treatment Costs Costs connected with the treatment of production process wastes are the least clearly defined costs of the three subcategories (treatment chemicals, testing, and disposal) mentioned earlier. Oftentimes, the same chemicals used in waste treatment are used elsewhere in the facility. For example, sodium hydroxide might be used to adjust pH in the wastewater treatment system and be used in a parts cleaning process. When collecting cost data for materials purchases, be sure to consider the amounts used in waste treatment as well.

There are several methods for determining usage of chemicals in waste treatment when separate purchasing records are not available. First, discuss treatment system operations with the department supervisor, who should know the quantities of each chemical used per week, month, quarter, and so on. Second, examine the wastewater treatment system itself and note where chemicals are added. Usage rates for wastewater treatment chemicals can be calculated from the dosage rates of metering pumps and the duration of treatment operations.

The third method of calculating waste treatment chemical usage involves back calculating theoretical chemical amounts given known flow rates and discharge limits. The chemical reactions involved in reducing waste toxicity can be used to determine the number of moles of each treatment chemical required. The number of moles of chemicals can be multiplied by the molar mass of the chemical and proportionally scaled up to determine total chemical usage. This is the least accurate of the three methods for determining total usage.

Laboratory Testing Costs Wastewater, hazardous wastes, and recyclable materials all require testing to verify concentrations. These tests are usually conducted by outside laboratories, which charge a fee based on the number of criteria being tested. The costs for these tests and their frequency should be available through the accounting department or on file with the wastewater treatment operator. If this information is not available through the accounting department, the costs can be found by contacting the lab directly. The environmental management department is required to have the results of hazardous waste and wastewater tests on file in case of inspection by state or federal regulators. The names of the laboratories conducting the tests and their telephone numbers would be listed on these reports or the cover page. A telephone call to the lab will yield a unit price for each test. For recycled materials, the hauler's invoice should state quantities of metal, plastic, or paper collected, and depending on the market price for these goods, the invoice may show a credit or an amount due.

Reviewing invoices from recyclers is another method to raise red flags on opportunities for pollution prevention. In this case, if the amount of material being recycled appears excessive (i.e., 50% of all material purchased ends up as scrap), further investigation into material use efficiencies are warranted.

Disposal Costs Disposal costs for wastes can be found on the invoices from the waste haulers. Telephone calls to the waste haulers will provide up-to-date unit prices for broad categories of wastes. Examine the hazardous waste manifests for the waste classification codes that correspond to these categories. Solid waste disposal costs are based on container size and frequency of pickup. Consult the waste hauler contract or a recent invoice (usually quarterly). Solid wastes are typically comingled, meaning that many different wastes from the entire facility are placed in the same container and thus share a portion of the total cost for removal. This portion can be determined two ways. One way is to conduct a "dumpster dive." The contents of a typical dumpster are unloaded and grouped into categories such as plastic, paper, food wastes, office equipment, and so on. Each group is weighed and computed as a percentage of the total. This percentage then can be used to determine which operations should share the costs of solid waste disposal (i.e., cost sharing). Aside from calculating cost share, a dumpster dive can reveal weaknesses in a facility's recycling programs.

For example, a company assumed all newspaper and corrugated cardboard were segregated and placed in separate containers for recycling. But on review of the dumpsters' contents, 50% of the volume was occupied by corrugated cardboard from the cafeteria. The workers in that area thought the cardboard was too dirty to be recycled and put it in the dumpster instead.

The second method of determining waste profiles involves collecting the waste before it enters the dumpster and measuring its volume. Cost share then will be proportioned by this volume as a percentage of total container volume.

Safety Training and Equipment (Hidden)

Federal environmental regulations, such as the community right-to-know laws and occupational safety and health regulations, require companies to train their workers about the risks involved with the materials used on-site and how to respond in the event of an accident. This training takes time away from production and has a quantifiable cost that can be allocated to each chemical and material used.

Example Calculation To illustrate, assume a company has 50 workers who are regularly exposed to the chemicals and require safety training. These workers come from any of the seven production centers in the plant and none comes in contact with all 50 chemicals. The safety training takes a full 8 hours to cover all material regarding the 50 chemicals. The average wage at the facility is $10.00/hr without benefits. Multiplying the number of employees at the training by their hourly wage and the total time for the training yields the annual cost to the company for this training event (50 employees × $10.00/hr × 8 hrs = $4,000). There also may be the cost of labor of the environmental or health and safety director, or a consultant's fee for performing the training.

If the number of chemicals involved or the number of people trained can be reduced, then the company saves money. Switching to nonhazardous chemicals in one or two of the production units could decrease the number of trainees. Replacing six chemicals that perform the same function with one similar or superior chemical also would decrease training time. Consult the person in charge of maintaining the Material Safety Data Sheets to obtain the total number of materials covered in the training and compare this with the number of those chemicals used at the process being analyzed. This ratio of process chemicals to total chemicals can be multiplied by the total training cost and used to allocate a portion of safety and health training to the process. Be aware that some of these chemicals may be used in additional locations. You may desire to subdivide these costs further to accurately represent usage of the chemical at other locations. Here is an area where a great deal of time can be spent on minutia. Use discretion and engineering judgment in deciding how to allocate the costs of these chemicals.

Document and consider chemicals that pose acute health risks or have known chronic toxic effects like cancer, birth defects, heart palpitations, and so on. The health and safety director should be able to identify these substances from the MSDSs. Most are labeled as carcinogens, teratogens, mutagens, radioactive, and so on. (See the Glossary for definitions of these terms.) The TCA offers a unique opportunity to bring these risks to the attention of management by footnoting these facts at the end of the analysis. If the project has a marginal net present value, this added information may be needed to make a more informed decision. Red flag the other high-risk chemicals and keep them in mind for future pollution prevention projects.

Personal Protective Equipment (Hidden)

Handling these chemicals often involves wearing protective equipment (face masks, respirators, gloves, aprons, boots) to protect workers from inhalation, splashes, and repeated contact with the chemicals (see Table 2.2).

In the electroplating industry, for example, line operators usually wear arm-length gloves, aprons, goggles, and sometimes boots to guard against contact with strong acids, caustics, and toxic metals. Purchase of this personal protective equipment (PPE) and the replacement of worn-out equipment should be allocated in the TCA. The facility health and safety coordinator should be able to furnish these costs from the annual budget. A project that eliminates the need for a hazardous chemical could also eliminate the need for PPE. This represents two cost savings to the company; the first cost of not buying the PPE or having to replace it, and the second larger cost of compensating a worker for an on-the-job injury or health effect incurred from using the chemical. Measuring the second cost savings ventures into subjective analysis and is not usually quantified, but may have impacts on the insurance cost category.

Sewer Usage Fees (Hidden)

Most industries are charged twice for their water use whether they recognize it or not. First, the company is charged for drawing on the town's municipal water supply. Second, the town charges the company for discharging the water into the sewer system. Exceptions to this rule are industries with their own well systems or intake from a river, direct discharge of effluent to

■ TABLE 2.2 List of Personal Protective Equipment Common in Industrial Processes

Respirators	Boots
Dust masks	Tyvek suits
Filter cartridges	Smocks or lab coats
Goggles	Hard hats
Face shields	Ear plugs
Approns	Hair nets
Gloves (many different kinds)	

streams or other water bodies (through a federal National Pollution Discharge Elimination System permit), and injection of wastewater into deep wells. In larger, older facilities (such as textile mills), tracing water sources and discharges can be a time-consuming but essential task. Not knowing where all the pipes lead can leave a company open to tremendous financial liability. A responsible environmental manager or engineer must exercise authority in this area as in previously mentioned areas (record keeping of waste classifications, determination of report filing responsibilities, and analysis of materials flow through processes.)

A compelling argument for undertaking a trace of water sources and discharges can be seen in the 1995 Q River spill from Bridgeport, Connecticut (see Chapter 9 for more detail on this event). Knowledge of where all pipes in a facility discharge is useful (and potentially job-saving), but not required to complete a TCA on a particular process. The best recourse during the TCA is to remain focused on the process in question and examine the supply and discharge pumps for that area. Knowing the source of waters or fluids coming into the pumps and leaving the process is the first step. Ratings on the pumps along with pipe diameters and hours of operation will generate flow volumes. These along with other measures such as supply or reserve tank capacities and the number of times these tanks are filled and emptied each shift can be used to generate water volumes and the sewer usage fee associated with that process.

Review bills in the accounting department or call the local sewer treatment authority to determine the sewer and water costs. Most towns issue one bill for water and sewer usage, and bill on a dollars per 100-ft^3 rate. Conversion is only necessary if the alternative process is not rated in the same units. Make sure to proportion the water and sewer costs accurately, especially if the process under consideration also has an in-line evaporator or closed-loop/zero-discharge system. In these cases, the water going in would be greater than the water leaving the process, and the sewer usage charge would not be equal to the water supply charge.

Regulatory Compliance (Hidden)

Environmental professionals understand the level of effort required to maintain compliance with today's environmental regulations. These responsibilities include

- filing for new or renewed permits
- sampling wastewater treatment effluent

- filling out weekly or monthly wastewater sample report forms
- conducting annual chemical and waste inventories and filing associated reports
- filling out hazardous waste manifests
- conducting weekly or daily inspections of hazardous waste storage areas
- keeping these inspection logs up to date
- providing annual training on hazardous materials and wastes
- updating emergency response, spill response, and contingency plans for the facility
- performing seasonal sampling of stormwater

Many states require additional reporting on chemical usage and pollution prevention activities. These tasks can add 100 to 400 additional hours to the existing workload, depending on the size of the facility and the number of chemicals used. The amount of time required to maintain compliance at even a medium-sized manufacturing plant facility can be more than a full-time job.

Pollution prevention has the potential to decrease the environmental manager's workload by eliminating hazardous chemicals and wastes or decreasing emissions from the facility. A reduction in the amount of hazardous waste generated also reduces the number of times the environmental manager has to fill out hazardous waste manifests. Elimination of a chemical regulated by a state's toxics-use-reduction law likewise decreases the amount of paperwork for reporting under this law. Industries can completely drop below the reporting threshold for a law and never have to perform the tasks connected to maintaining compliance with that law. One manufacturer implemented so many pollution prevention projects at its facility that it changed the Resource Conservation and Recovery Act (RCRA) waste status from Large Quantity Generator to Very Small Quantity Generator. Aside from saving money on hazardous-waste disposal, this status change eliminated the need to perform weekly hazardous-waste inspections, keep logs of these inspections, maintain a hazardous waste contingency plan, and train workers on how to respond to a hazardous-waste spill. Best of all, the change in status guaranteed that the state Department of Environmental Protection would never show up at the facility to conduct a hazardous-waste inspection or cite for a violation. The manufacturer could make this claim because the state Department of Environmental Protection spends all of its time inspecting the Large and Small Quantity Generators and leaves the Very Small Quantity Generators alone. Potential savings like these must be captured for the TCA.

How to Capture the Paperwork Savings The cost of filing paperwork for ongoing monitoring is easily obtained by estimating the amount of time to complete the task once and multiplying by the number of times the task is performed per year. To illustrate, suppose an environmental manager spends 15 min/week inspecting the hazardous waste storage for labels on the drums, noting collection dates, possible mixing of incompatible wastes, looking for signs of leaks or damage to the containment berm and floor, and completing the inspection log. This inspection is performed 50 times per year and accounts for (50 × 15 minutes/60 min/hr =) 12.5 hr. If the manager/engineer earns $20/hr, this task costs the company $250 per year. If the process being analyzed is the only generator of hazardous waste in the facility, then the entire amount is entered. Otherwise, the total of all tasks can be split among all hazardous-waste-generating activities.

Another approach is to allocate a percentage of the environmental managers total time to specific processes such as the wastewater treatment system or the baghouse dust collectors. An estimate is made based on the percentage of time spent on environmental compliance issues related to each waste treatment area. If checking the wastewater treatment system and sampling its effluent takes half of the managers time, then 50% of the manager's salary would be allocated to this system. The total effluent from this system then would be apportioned to each process that discharges to it. A water curtain behind a spray painting line may contribute 25% of the total flow through the wastewater treatment system. Therefore, paperwork costs for the painting line can be estimated as:

Environmental manager's salary ($40,000)
× 50% of time spent on wastewater treatment system
× 25% of effluent coming from spray painting processes
= $5,000

A similar calculation can be performed for each waste stream leaving the process. The total of these creates the total cost of all paperwork connected with the process. Obviously, this method is limited by the ability of the manager/engineer to estimate and allocate his or her time.

If these steps seem to be too much work to obtain the numbers, consider using some of the formulas and average values provided in the EPA *Pollution Prevention Benefits Manual.*[3] Several of these formulas and averages are provided in Appendix B for reference.

[3] U.S. *Pollution Prevention Benefits Manual*, U.S. EPA Office of Policy, Planning, and Evaluation, EPA230/R-891/100, Washington, DC: 1989, pp. 3.11–3.25.

Nonannual Events

Many environmental compliance events occur only once every 2, 3, or 5 years, such as the renewal of a wastewater treatment discharge permit, an update to a pollution prevention or waste minimization plan or preparation of biennial hazardous-waste-reports. The costs for such nonannual events can be annualized by dividing the number of years into the total cost. Annualizing these costs does introduce a degree of error into the TCA, because the money is spent as a lump sum every 2 years, not in equal annual amounts across time. Taking the present value of an annualized cash flow versus a periodic cash flow results in a difference in the total present values of the event. The following example demonstrates this difference between the present values of the annualized cost for a $2,000 Massachusetts Toxics Use Reduction Act (TURA) plan update (prepared every 2 years for 6 years) and the cost when incurred.

Year Cost Is Incurred	Present Value of Annualized Cost	Present Value of Cost When Incurred
0	0	0
1	$909	0
2	$827	$1,653
3	$751	0
4	$683	$1,366
5	$621	0
6	$565	$1,290
Total present value	$4,356	$4,309

This example shows a difference of $47 between the two methods. Annualizing this periodic cost created a 1.1% overestimate of the present value for the cost of the TURA plan updates. Because this is a cost and not a savings, the overestimate has the benefit of making the analysis slightly more conservative with regard to the amount of money that will be saved. Perform this allocation carefully to prevent double counting if more than one project will impact these types of costs.

Waste or Chemical Use Taxes (Hidden)

Many states charge a tax for the generation of hazardous wastes; Connecticut, for example, charges $0.005/lb. If a Connecticut industry generated

50 tons of metal hydroxide sludge per year, it would owe the state $500. These taxes fund programs to reduce the amount of hazardous waste generated in the state, or assist in cleaning up spills or abandoned industrial properties. Other states have fees or taxes based on chemical usage and/or company size. Massachusetts charges $1,100/year per listed toxic chemical and a second fee based on the number of full-time employees.

If the process under consideration is also the one generating the waste or using the taxed chemical, there could be costs here that need to be incorporated. In some cases, the elimination of the chemical saves not only the purchase and disposal costs, but also the usage fee and waste tax. The accounting department should have receipts from checks sent to the state or local governments. Also check the files containing state-mandated reports or pollution prevention plans. Copies of checks and certified mail stubs should be attached to the file copies of these reports. Care must be taken not to double count these fees or taxes. This tax must be allocated according to the number of chemicals used in the process that are subject to the tax or fee.

Insurance (Hidden)

All manufacturing companies are required by law to carry a variety of insurance coverage. These premiums fall into several broad categories:

- Health insurance and workers compensation for employees
- Fire, theft, and damage for properties
- Liability insurance against accidents and intentional harm
- Partial or total contingent liability coverage against pollution exposure

Companies pay for these insurance premiums out of their operating budgets and rarely allocate the insurance cost to operations or events covered by the insurance. Information about premiums should be available through the accounting or legal department of the company. Of the coverage listed, only the rates for workers compensation are currently impacted by prevention of incidents.

One example of this can be seen at a facility that instituted several pollution prevention measures that also reduced muscle sprains and back problems for the workers. These improvements contributed to a decline in on-the-job accidents. The company instituted a program to reward workers to take care and avoid accidents. The company's insurance agent was able to reduce the premium cost for workers compensation coverage because of the decrease in accident claims. A portion of the savings from the reduced insurance premiums was given away in a lottery to all workers who had not had accidents that year.

Insurance companies that do cover "pollution exposure" claim to consider all potential risk before they set premiums. However, there is only anecdotal evidence to suggest that premiums would be reduced if pollution exposure risk was decreased (see Chapter 9 for additional detail on commercial liability insurance).

The effects of insurance are considered here in hopes of encouraging insurance carriers to look more closely at the benefits of pollution prevention. Clearly, by eliminating a chemical that could cause an explosion, or poison the community, a company has reduced the probability that the insurance company will have to pay a claim. In a just world, this would be reflected in the insurance premiums. To date, there are no known direct reductions in insurance premiums as a result of pollution prevention efforts. States such as Illinois are beginning to educate insurance providers about the benefits of pollution prevention, which hopefully will initiate changes in policies.

The examples and case studies in this book retain the insurance cost category as a reminder to explore this potential cost savings. Managers and engineers are encouraged to retain this category in the TCA to initiate discussion among management about the potential for premium reductions. As more insurance companies learn of the loss reduction benefits of pollution prevention, they will begin incorporating them in their policies.

Production Costs (Hidden)

Production costs are considered hidden costs because oftentimes a change in manufacturing process will have a positive or negative impact on the rate of production. Think of production as the rate at which materials are converted into product. A change to one step in the manufacturing process may have a cascade effect all the way down the line. Refer to Eli Goldratt's *The Goal*[4] for more information on the theory of constraints (a short summary relating to TCA is provided in Chapter 10). In brief, no system can operate at a greater capacity than its slowest process. Therefore, a change in any step in the manufacturing process can have one of two outcomes: It can either increase or decrease the capacity of the slowest process, or increase or decrease the capacity of some other process step. In the former case, production, or

[4] Eliyahu Goldratt, *The Goal*. Great Barrington, MA: North River Press, 1992, p. 67.

throughput, will also increase or decrease; in the latter, throughput will remain constant or decrease if this other process step now becomes the slowest process. Obviously, there are costs involved in lost production capacity. If the change in the process allows other functions to occur simultaneously, there may be an offset to the decrease, resulting in zero gain or loss.

For example, a machine shop replaced its vapor degreaser with an aqueous parts cleaning system. On the old vapor degreaser, the operator had to hold the basket in the degreaser. This was not an issue because the parts were cleaned relatively quickly and a single employee could be dedicated to this task. The aqueous system required a longer cleaning time, but allowed the operator to walk away from the process and perform another task. The loss in production time due to a longer cleaning process was compensated for by freeing the employee to perform a simultaneous task.

The production cost will be only applicable to the option being considered, as each project may effect the cost of production separately and that cost may change due to factors outside the scope of the analysis. Increases in union labor rates, import export taxes on raw materials, and utility rate increases will all affect production costs and change independently from the impacts of new technology. Quantifying the current production cost is not as important as measuring the increases or decreases due to the proposed project. These costs or savings are more readily determined from known manufacturers' specifications of feed rates, processing or drying times, and other variables usually detailed in sales brochures.

Theoretical or actual time/motion studies can be performed to determine work flow rate and potential increases or decreases. A time-and-motion study reviews all the steps in a process and records the time required to complete each one. Counts of parts processed per hour or time required to process one part then can be developed. Changing one step, for example, extending the dwell time of a rack of parts, can be shown to increase the total processing time. This increase then can be extrapolated out for an entire year's operation and the total increase or decrease in throughput determined. If the change in the single step is more labor-intensive, the total amount of additional labor required can be calculated. If the change eliminates one or more process steps, this effectively creates extra production capacity as more product can be processed in the same amount of time. Extra production capacity can be used to fine-tune other process steps. For example, an extra 7 seconds of capacity on an automatic electroplating line could be distributed among programmable hoist instructions to slow rack withdrawal rates on the major electroplating tanks. This would cause more plating solution to remain in the plating tank, requiring less rinsing and creating less wastewater treatment sludge.

The impacts of changes in production over the period of 10 years (typical project lifetime) can add considerable value to the project. A $10,000 decrease in production costs has a present value of $61,400 over 10 years, not including increased revenues from sales of goods produced with the extra capacity.

Replacement and Closure Costs (Hidden)

Industrial facilities that have existed on the same property for decades are faced with replacing or extracting underground fuel storage tanks. Federal and state regulations require tightness tests, record keeping, and eventual removal of aged tanks before they begin to leak. The manager/engineer has to examine these current or future costs and determine if they should be allocated to the process under analysis. These costs should be included if the process is the sole or major user of the fuel oil. Changes to this process could increase or decrease the need for the fuel oil tanks. A change in fuel source (e.g., to natural gas) may eliminate the need for these storage tanks, resulting in both costs and savings.

Potential costs would include

- draining and extraction of underground storage tanks
- cleaning and decommissioning of above-ground storage tanks
- closure and capping of all piping
- disposal of the tank structure and the collected wastes
- removal or remediation of any surface or subsurface contamination

Potential savings would include

- decreased need for daily, weekly, or monthly inspection of fuel levels (labor)
- decreased record keeping requirements (paperwork)
- decreased long-term liability for soil contamination

THE ISSUE OF MISSING DATA

In a perfect world, there would be "perfect data," that is, everything you needed to know about the process in question would be available, accurate, and in the proper units. In reality, a company keeping exceptional books and maintaining its equipment well will still have only 70% of the data required. As mentioned earlier, very few companies expend the energy required to monitor and record cost data at the fine-grained level of detail needed for a TCA. And this is okay. If everyone spent the time required to collect every scrap of data available, no one would ever actually produce anything.

Accounting systems and production systems seek their own equilibrium in response to the daily demands of the market, management, labor, and external influences such as the federal regulations, the national economy, and the weather. If there was a demand for the process-level data (such as is created by ISO 9000 certification), the company's accounting and production systems would create methods to collect the data. But for most companies, conducting a TCA will be their first introduction into collecting process-level data. Because of this, the manager/engineer should expect that at least 30% of the data needed will never have been recorded before.

METHODS TO COMPENSATE FOR MISSING DATA

This lack of data will necessitate some creativity in generating numbers for the analysis. Several approaches exist to plug these data "holes."

Method 1. Estimate Based on Similar Processes within the Facility

A subprocess or smaller specialty process may use identical chemicals as a large process. Chemical use data for this small process may not exist because both processes are considered by the accounting or purchasing departments to be one unit. Oftentimes, it is not known exactly how much material is used on these smaller processes. Materials needs are met through the contingency quantity tacked on by the plant manager during requisitioning or by the purchasing department during ordering. This contingency amount can be as much as 10% of the total order. Extra material is ordered to ensure that production will have enough, rather than run out, and cause a production delay or workflow stoppage.

Example of Method 1 A line operator for a small process borrows material stock from a similar larger process to meet the small demand, and keeps no record of the amount borrowed. How can this amount be determined?

If the smaller process is similar in operation and material input, it may be possible to apply the ratios of material used in the large process to estimate quantities used in the small process. For example, a pigment manufacturing company had several batch mixing machines. Among other additives, inputs of 100 kg of calcium and 50 kg of magnesium (a 2:1 ratio) were combined in a large-batch mixing operation. The smaller mixing process also required calcium and magnesium in a 2:1 ratio, and several other additives not found in the larger batches. If the large process created 2,500 lb of pigments per batch and the smaller process created 500 lb of pigments per batch, then the smaller mixer used five times less ingredients per batch. Simple division yields an estimated input quantity of 20 kg of calcium and 10 kg of magnesium. There are obvious flaws to this method, some of which can be countered by carefully interviewing line operators to find similarities between the two processes and the products they produce.

Method 2. Estimate Based on End-Product Composition

If the chemical formulation of a mixture, alloy, or other composite is known, then the input quantities can be derived by multiplying the molar mass ratios with total annual units produced. Use the molecular weights of both the end product's formulation and the input material in question to calculate the mole and mass fractions. Then multiply the mass fraction by the total number of units produced. This yields the amount in weight of input material used to create all the units produced in 1 year. All that is required now is a unit price of the input material or chemical, which can be obtained from invoices or directly from the material supplier.

Example of Method 2 Suppose a company creates deicing pellets (calcium chloride) from limestone (calcium carbonate) by dissolving the limestone in hydrochloric acid. The company simply obtains limestone rocks from its property and combine it with a by-product acid in a tank. The chemical reactions would look like this:

$$CaCO_3 + 2HCl \rightarrow CaCl_2 + H_2O + CO_2$$

The company uses waste hydrochloric acid and would like to know how much acid it put in but there are no input records. How can the quantities be derived?

The manager/engineer knows from the equation that 2 mol of hydrochloric acid are needed to create 1 mol of calcium chloride. The company knows that its bagging machine weighed out one thousand 20-kg bags of the deicing salts. Of this 20,000 kg of salts, 90% is pure calcium chloride. From the periodic table of elements, the molecular weight of calcium chloride is

- Ca molecule (molecular weight of 40 g/mol) and
- chloride molecules (molecular weight of 35 g/mol each)

or 110 g/mol. To find the number of moles of calcium chloride produced, we take 90% of 20,000 kg, or 18,000 kg and divide by 110 g/mol. After converting units, we have 163,636 mol of calcium chloride. From the earlier formula, we know that the reaction ratio of HCl to $CaCl_2$ is 2:1, therefore we need at least 327,272 mol of hydrochloric acid. Next we multiply this by 36 g/mol (the molecular weight of HCl) to get 11,781,792 g (or 11,782 kg) of pure hydrochloric acid. Most acids come as a liquid though, so the final step in this calculation is to divide by the acid's specific gravity. Because acids can be purchased in a variety of strengths, the manager/engineer would have to consult the Material Safety Data Sheet on file at the facility for the proper specific gravity for this blend of acid and also account for the dilution of the pure acid to working strength. For this example, we will use a specific gravity of 1.1513 kg/l at 30% by weight and 20°C. Divide 11,782 kg by 30% to obtain the total weight of the diluted solution, and then divide this by the specific gravity given before to obtain 45,215 l of hydrochloric acid, or roughly 11,946 gal.

The method just detailed does not account for improper mixing of solutions, spills, or the effects of impurities and temperature. It is a theoretical value and best treated as a minimum value of acid required to derive the total product. The example does illustrate the potential to derive information using basic chemistry principles.

Method 3. Conduct a Survey or Benchmark

It is likely that the company has competitors with similar processes. Industry trade groups and research organizations can provide generic process information on material inputs and rates. Basic input/output ratios and materials feed rates are often provided in trade literature, help columns of monthly trade magazines, and reference books. This approach is similar to what is stated earlier with information coming from outside the facility rather than inside. Keep in mind that this data will be generalized and should be used only as a reference point. Always footnote to describe where the data came from and how they may or may not represent operations at the facility.

Method 4. Take an Educated Guess

In the interest of using time efficiently, draw on the resources of your staff and the experience of the plant managers and foremen. These individuals work with the process every day and have had to troubleshoot on their own when the process goes down. They will know the process inside and out. Although they may have never measured the inputs and outputs, they could offer an educated guess.

This sort of guesswork was common practice within the textile industry. The line operators formulated the process baths using shovels instead of measuring cups. They would add a shovel or two of this and a shovel or two of that and produce the desired results. If it was slightly off, these people would know what to add to correct it. They never bothered to measure or write down the formulations; they just knew through on-the-job experience.

These people will answer the question by describing the methods used. It will be the manager/engineer's responsibility to derive mass and material quantities from the description of their operations. The derivation may require measuring the volume of an average shovel load, for example.

Method 5. Leave It Blank

It may be appropriate to leave certain cost categories blank, either to minimize the amount of time spent on the TCA or to limit inaccuracies in the calculations. The manager/engineer should footnote any blank category to explain why no data were entered and the possible effect this will have on the analysis. Always keep in mind that TCA is a financial decision-making tool, not an abstract theoretical exercise. It is very easy to get lost in the derivation of numbers. Engineers are especially fond of calculating theoretical yields and energy requirements. This is not the main focus of TCA! *The main focus of TCA is to use cost accounting to show the financial merits of environmental projects, especially pollution prevention projects.*

SPEED VERSUS ACCURACY

Leaving cost categories blank is most noticeable when using the shortcut method of performing a TCA (see Chapter 5). The shortcut approach examines the largest cost categories first. Then, if the project has a marginal present value, other cost categories are introduced to clarify the financial feasibility. This method saves time by looking at the largest cost categories and leaving all others blank until, and if, they are needed.

How does this approach impact data gathering? If one goes into the TCA with a commitment to look at just the large cost categories, there is a high probability the data will already exist and not need to be derived. Having all the data available will dramatically decrease the amount of time required to perform the TCA. Expediency can be an advantage if a large number of analyses must be performed or a last minute request for an analysis is made. Having a quick method to perform a TCA allows the original cost savings projections to be updated as the project is implemented or as conditions change (shifts in market prices for materials, labor rates, etc.). Having a quick method for TCA provides a way for the manager/engineer to track the progress of the project and generate proof that it is saving the company money.

What Is Considered the Most Significant Cost Category?

For each facility, and each process within that facility, the major cost categories will be slightly different. For example, production costs can be the most significant cost category when automating technologies are installed. In the case of adding a second hoist to an electroplating line, production capacity on that line almost doubles, effectively halving plating time. Yet this improvement also allows extended dwell times over each tank to reduce solution loss and significantly decrease hazardous-waste generation. In this case, production cost and waste disposal would be major cost categories. In the example mentioned earlier of the electroplater adding a fourth plating line, storage costs can become a major cost category when floor space is scarce.

In general, chemical purchases, waste disposal, and utilities make up the majority of costs connected to any one process. Maintenance and production impacts are the next largest, followed by labor for paperwork, sewer usage fees and taxes, and safety training/equipment. At a facility level, wages, benefits, and taxes are the largest cost categories, with materials costs closely trailing.

You must strike a balance between credible financial data and time-efficient analysis of projects. If all the categories were filled in with inaccurate estimates, the project, at worst, may appear too profitable. Leaving too many categories blank may cause the project to be discarded for not achieving the required return on investment (ROI). Short-selling pollution prevention costs a company money in the long run and is counterintuitive to doing what is recognized as the best approach to environmental management.

As manager and evaluator, you must find the level of detail with which you are comfortable operating. Perhaps in the first several TCAs, the data are simply not available and many cost categories are blank. Blank spots should send red flags to the facility's environmental management team to review the cost accounting and materials tracking systems. Changes to these systems do not have to be companywide, although they are best implemented from the top down. If that is not possible, working with line operators and foremen to log usage on specific processes and regular calls to the purchasing or accounting departments for unit costs are adequate substitutes. These improvements will allow detailed analyses that are easier to perform and more accurate. Smaller companies have an advantage here, especially when the president is much closer to the line operators or is handling the environmental issues. Changes in policy will flow quickly and directly in this case. As is described in Chapters 5 and 6, a company can customize formulas and cost categories for their facility and processes that will make TCA as easy to perform, or easier, than a payback analysis.

PURCHASING DECISIONS AND DATA GATHERING

Recognized or not, the purchasing department of any corporation exerts enormous influence over the environmental risks of the company. Purchasing agents are too often influenced by vendors and have never been given the benefits of sound environmental information. They are often inundated with "free samples," which then have to be disposed of at company expense.

Educating the purchasing department about environmental issues is one method to decrease the number of TCAs that will have to be performed in the future. An environmentally educated purchasing agent will know not to order CFCs, will ask questions about worker risk from new chemicals, and will be suspicious of "miracle solutions." Eliminating the purchase of hazardous materials from the facility directly eliminates risk and liability. Although not directly related to TCA, the purchasing process is the front line in any company where decisions on materials can be influenced. Oftentimes, purchasing decisions are based solely on least initial cost without regard to the cost of waste disposal, environmental risk, or worker health issues. Preventing the purchase of a toxic material will eliminate the need to find an alternative farther down the road.

CASE STUDIES OF DATA GATHERING

The following two case studies present situations encountered at two facilities where data gathering became particularly challenging. Careful study will reveal subtleties in each that point up valuable lessons and useful "tricks of the trade." The first case study details a materials mass balance conducted at a leather tannery. Here, there was very little information available about the processes and materials used. The second case study presents a very basic accounting of operating costs for a leather dyeing operation in which very little cost data were available.

CASE STUDY 1: Leather Tanning

A medium-sized leather tannery desires to know how much chromium is used in the processing of its hides. This case study will discuss the methods used to gather materials use data and costs for the chromium compounds. It offers direct process measurements as primary data sources, as well as estimation methods for determining missing values.

Step 1: Process Overview

Raw animal hides, or "blue stock" is shipped to the tannery from Midwest cattle houses. Raw animal hides require unhairing and tanning, whereas blue stock is pretanned. The poor availability of quality hair hides during the winter forces the use of mostly blue stock and a subsequent decrease in tanning operations during 3 months of the year. Hair hides are unhaired and tanned through batch processing in (four) 25,000-gal tanks. A mixture of sodium sulfide and sodium sulfhydrate is used as a wash and unhairing solution. Hides are rinsed and then tanned in the same tanks using a chromium solution. At this point, the hides are now considered "blue stock" and leave this production unit as wet, chromium-containing, tanned hides.

Blue stock enters the sort, splitting, and shaving process resulting in a "top-grain" leather, a first split, second in quality to top-grain, and a second split, usually considered a waste product. This process also generates scraps and dust that contain chromium. The blue stock is sent to the coloring department, where the blue sides are chemically processed with dyes, syntans, extracts, and fat liquors to produce a final product. Batches of 150 sides of blue stock are placed in wheels and tumbled in dye solutions. The dye cycle follows with a rinse cycle and unloading of sides. These are then pasted to a drying frame to stretch the sides and eliminate wrinkles. Dyed and dried sides are then brought to stakers and dry mills to impart softness and flexibility characteristics. This process generates fine dust that contains chromium. Hides then may be painted or embossed with a pattern to simulate various animal skins, such as crocodile and ostrich. Figure 2.1 shows the general process flow for the tanning operation.

Step 2: Quantifying Inputs and Outputs

To determine the quantities of chromium used and lost during tanning, several samples of tanned hides at various stages of production were taken. Samples of wastes and rinsewaters were also analyzed for chromium. The following are the sources and losses of chromium in the tanning process and a description of the calculations and assumptions made.

1. *Chrome Used in Tan Tanks.* The company's 1996 chemical inventory records show the following purchases of liquid chrome tanning solution:

1996 Begin inventory	640 gal
Chrome solution No. 1	42,734 gal
Chrome solution No. 2	2,890 gal
1993 End inventory	1,000 gal
Total 1993 chrome solution usage	45,264 gal
Specific gravity of this solution (from MSDS) = 12.91 lb/gal	
Total weight of chrome solution used	584,358 lb

Material Safety Data Sheets and conversations with chemical distributors reveal that tanning solution Chrome 42A is delivered to the company at a concentration of chromium oxide of 11.75%. Chromium molecules make up only 26% of the total molecular weight of chromium oxide. Therefore, the actual usage of elemental chromium is

$$584{,}358 \text{ lb} \times 11.75\% \times 26\% = 17{,}852 \text{ lb Cr used in 1993}$$

The company recorded 109 soak days (tans/soak day) during 1993. This averages out to 207 gal of tanning solution per tan batch (lower

■ FIGURE 2.1 Process Flow Diagram for Leather Tanning

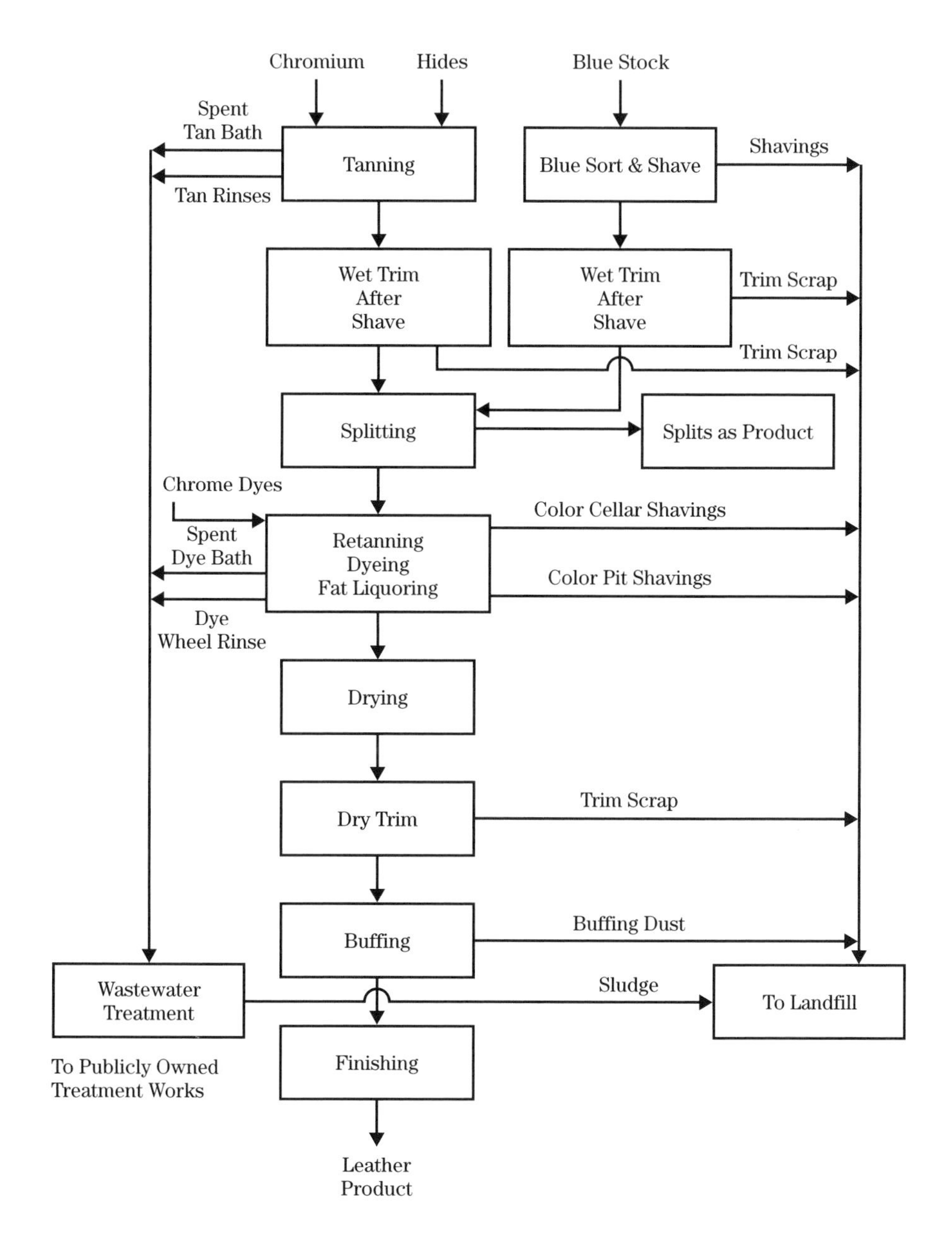

than the specified 215 gal). Each tanning batch (for 1996), therefore, contained an average of 82 lb. of chromium metal.

2. *Chrome Lost in Spent Tan Solution.* Spent tanning solutions are drained from the turbo tan tanks after tanning is complete. The company staff sampled the spent solutions and had them analyzed for total chromium content.

Sample No. 002 contained 2,710 ppm chromium.

During 1996, the average volume of tanning solution in each tank was 207 gal. The specific gravity of the solution is 12.91 lb/gal. Therefore, the residual chromium in the spent bath is:

$$2{,}710 \text{ ppm} \times \frac{1}{1{,}000{,}000} \times 207 \text{ gal.} \times 12.91 \text{ lb/gal} = 7.24 \text{ lb Cr/tank}$$

$$7.24 \text{ lb Cr/tan} \times 2 \text{ tan/day} \times 109 \text{ days/year} = 1{,}578 \text{ lb Cr/year}$$

3. *Chrome Lost in Tank Rinse.* The Turbo Tan tanks are rinsed with 600 gal of water following the draining of the tan solution. This rinsate was sampled and analyzed for total chromium. The specific gravity of water is 8.2346 lb/gal.

Sample No. 008 contained 2,150 ppm chromium.

$$2{,}150 \text{ ppm} \times \frac{1}{1{,}000{,}000} \times 600 \text{ gal/tank} \times 8.2346 \text{ lb/gal} = 10.6 \text{ lb Cr/rinse}$$

$$10.6 \text{ lb Cr/rinse} \times 2 \text{ tans/day} \times 109 \text{ soak days/year} = 2{,}315 \text{ lb Cr/year}$$

4. *Chrome Transferred to Wet Splits.* The company sampled the tanned leather as it exited the Turbo Tan tank and before it was wrung out. Testimony from specialists at the Leather Institute of America's (LIA) Research Laboratory in the University of Cincinnati, Ohio, revealed that the moisture content of hides at this process step could be as high as 80 to 90%. This must be accounted for in the calculations for total chromium in wet splits because the analysis numbers are based on dried samples.

Sample No. 001 contained 25,700 ppm chromium.

The company production records revealed the total number of hides processed in the Turbo Tans to be 72,505 for 1996. Given 109 soak days, 2 tans/day, and an average hide weight of 48 lb, this yields an average batch weight of 15,964 lb per tan. This is less than the 17,000 lb specified.

$$25{,}700\text{ ppm} \times \frac{1}{1{,}000{,}000} \times 15{,}964\text{ lb/tan} \times 15\%\text{ solids} = 61.5\text{ lb Cr/tan}$$

$$61.5\text{ lb Cr/tan} \times 2\text{ tans/day} \times 109\text{ days/year} = 13{,}407\text{ lb Cr/year}$$

This represents a 75% chromium adhesion efficiency prior to wringing.

5. *Chrome in Wet Trim before Wringing.* Tanned hides are trimmed before they are put into the wringer. Scrap generation data from 1988 was used as the starting point for these calculations. The assumption was that chromium concentrations in tan hides at this process step would be similar to those tested right after tanning. The moisture content was also assumed to be similar.

$$250\text{ lb scrap/day} \times 109\text{ soak days} \times 15\%\text{ solids} \times 25{,}700\text{ ppm} \times \frac{1}{1{,}000{,}000}$$
$$= 105\text{ lb Cr/year}$$

6. *Chrome in Blue Sides Received.* The company bought 223,160 pretanned sides at 24 lb/side = 5,355,840 lb of blue stock. Testimony from specialists at LIA Research Lab revealed that the average moisture content of blue stock after shipping to be in the range of 50 to 60%. The company staff sampled incoming blue stock.

Sample No. 003 contained 25,000 ppm chromium.

$$25{,}000\text{ ppm Cr} \times \frac{1}{1{,}000{,}000} \times 5{,}355{,}840\text{ lb} \times 45\%\text{ solid} = 60{,}253\text{ lb Cr/year}$$

7. *Chrome Loss in Blue Sort Shavings.* Blue stock is sorted on arrival and shaved. This generates scrap hide pieces that contain chromium. Scrap generation rates from 1988 were used as a starting point for this calculation. The moisture content was 65% (35% solids), as sides are wetted down to increase flexibility. Chrome concentration is the same as on arrival (25,000 ppm).

$$25{,}000\text{ ppm} \times \frac{1}{1{,}000{,}000} \times 35\% \times 223{,}160\text{ sides/year} \times 0.9\text{ lb scrap/side}$$
$$= 1{,}757\text{ lb Cr}$$

8. *Wet Trim after Shave Losses.* Blue stock is then trimmed further prior to coloring. Scrap generation rates were taken from 1988 data. The moisture content is the same as for the blue sort shaving calculation (35% solids). Chrome concentration also remains the same (25,000 ppm).

$$223{,}160 \text{ sides/year} \times 25{,}000 \text{ ppm} \times \frac{1}{1{,}000{,}000} \times 50 \text{ lb scrap/140 sides}$$
$$\times\ 35\% \text{ solids} = 697 \text{ lb Cr}$$

9. *Chromium in Blue and Black Dyes.* Table 2.3 records the two leather dyes used at the facility that contain chromium. Material Safety Data Sheets and conversations with the chemical vendor confirmed the concentration of the dyes and total percentage of chromium metal.

Blue and black dye runs contributed 755 lb of chromium to the input side of the materials balance equation in 1996.

10. *Chrome Lost from Spent Dye Baths.* Records from the 1996 Massachusetts Form S show 343,064 sides were dyed in the color cellar in 1996. Conversations with the plant manager revealed that 80% of these sides were dyed black and 90% of black sides contained chromium dyes. The plant manager further related that on average, 150 sides constitute a "color pack" and 7,500 gal is considered a good float for that pack. The company sampled the spent dye baths and had them analyzed for total chromium content.

Sample No. 005 contained 15.7 ppm chromium.

$$343{,}064 \text{ sides/year} \times 80\% \text{ dyed black} \times 90\% \text{ dyed Cr black} \times \frac{1 \text{ wheel}}{150 \text{ sides}}$$
$$\times\ 7{,}500 \text{ gal/wheel} \times \frac{15.7 \text{ Cr}}{1{,}000{,}000} \times 8.2346 \text{ lb/gal} = 1{,}597 \text{ lb Cr}$$

11. *Chrome Lost in Color Rinsing.* After a complete dye cycle, the sides are rinsed with water. All of the same parameters mentioned in calculation 10 apply to this calculation. The company sampled this rinsate for total chromium.

Sample No. 009 contained 5.58 ppm chromium.

$$343{,}064 \text{ sides/year} \times 80\% \text{ dyed black} \times 90\% \text{ dyed Cr black} \times \frac{1 \text{ wheel}}{150 \text{ sides}}$$
$$\times\ 7{,}500 \text{ gal/wheel} \times \frac{5.58 \text{ Cr}}{1{,}000{,}000} \times 8.2346 \text{ lb/gal} = 567 \text{ lb Cr}$$

12. *Shavings from the Color Cellar Screens.* The amount of shavings caught in the color cellar screens depends on the number of days the cellar runs. Data from 1988 formed the basis of this calculation, where an estimated 450 lb of shavings were generated in 1 day of operation. Year 1996 saw 228 operating days. The moisture content was taken to be 85% (15% solids). No samples of these shavings were taken, and the 2.38% Cr figure from 1988 was used.

$$450 \text{ lb of shavings/day} \times 228 \text{ dye days/year} \times 2.38\% \text{ Cr} \times 15\% \text{ solids} = 366 \text{ lb Cr/year}$$

13. *Shavings from the Color Cellar Pit.* Shavings are also dredged from the settling pit in the color cellar. No samples were taken from this area and 1988 data were used in the calculations.

$$200 \text{ lb sludge/day} \times 228 \text{ dye days/year} \times 2.38\% \text{ Cr} \times 15\% \text{ solids} = 163 \text{ lb Cr/year}$$

14. *Buffing Dust Losses.* The company sampled the buffing dust for chromium content. Calculations were based on these results and dust generation factors developed by the company in 1988.

Sample No. 005 contained 25,800 ppm chromium.

$$25{,}800 \text{ ppm} \times \frac{1}{1{,}000{,}000} \times 1 \text{ barrel of dust /day} \times 85\% \text{ solids} \times 55 \text{ lb/bl} \times 252 \text{ days/year} = 304 \text{ lb Cr/year}$$

15. *Dry Trim Chrome Losses.* Both tanned stock and blue stock is trimmed after it is dried to 15% moisture (85% solids). Scrap generation rates were taken from the 1988 data, and chrome concentrations were assumed to be the same as in the buffing dust samples (25,800 ppm).

■ TABLE 2.3 Chrome-Containing Dyes

Dye Name	% Chromium	1993 Usage (lb)	Weight of Chromium (lb/year)
Blue 2G 250	80	100	80
Black N-WA	5	13,490	675
Totals	N/A	13,590	755

Concentrations here are higher than incoming stock due to the addition of chrome-bearing dyes.

$$25{,}800 \text{ ppm} \times \frac{1}{1{,}000{,}000} \times 368{,}170 \text{ sides/year} \times \frac{170 \text{ lb scrap}}{1400 \text{ sides}} \times 85\% \text{ solids}$$

$$= 980 \text{ lb Cr/year}$$

16. *Chrome-Dyed Finished Product.* As mentioned before, approximately 72% (90% of 80%) of the product line is chrome-dyed black leather. In 1996, the company shipped 6,637,869 ft^2, of product. The average side measures to 20 ft^2, resulting in 331,893 sides of finished product. The moisture content is 15% (85% solids). The company sampled chromium content in both dyed and nonchromium-dyed finished goods.

Sample No. 006 contained 20,100 ppm chromium (chrome-dyed product).

Sample No. 007 contained 24,400 ppm chromium (nonchrome-dyed product).

(a) *Chrome-Dyed Product*

$$\frac{20{,}100 \text{ Cr}}{1{,}000{,}000} \times 72\% \text{ of product line} \times 331{,}893 \text{ sides} \times 85\% \text{ solids} \times 4.5 \text{ lb/side}$$

$$= 18{,}372 \text{ lb Cr/year}$$

(b) *Nonchrome-Dyed Product*

$$\frac{24{,}400 \text{ Cr}}{1{,}000{,}000} \times 28\% \text{ of product line} \times 331{,}893 \text{ sides} \times 85\% \text{ solids} \times 4.5 \text{lb/side}$$

$$= 8{,}673 \text{ lb Cr/year}$$

The company also sells splits. In 1996, 345,956 splits were shipped, at approximately 13 lb per split, with a chromium content similar to incoming blue stock (25,000 ppm).

(c) $2.5\% \text{ Cr} \times 345{,}953 \text{ splits} \times 35\% \text{ solids} \times 13 \text{ lb/split}$

$$= 39{,}352 \text{ lb Cr/year}$$

Summary of Inputs and Losses

Table 2.4 summarizes chromium inputs and outputs for the company's tanning and finishing processes.

These calculations account for 97% of all chromium used on-site at the company. Possible areas not accounted for are chrome-bearing

■ TABLE 2.4 Mass-Balance Summary

Chromium Source	Input (lb/yr)	Output (lb/yr)
Tanning solution	17,852	
Blue stock received	60,253	
Chrome dyes	775	
Spent tan baths		1,578
Tan tank rinse		2,315
Blue short shavings		1,757
Wet trim after shave		697
Wet trim before wringing		105
Spent dye bath		1,597
Color wheel rinse		567
Color cellar shavings		366
Color pit shavings		163
Dry trim		980
Buffing dust		304
Chrome-dyed product		18,372
Nonchrome-dyed product		8,673
Splits		39,352
Total	78,860	76,826

solutions wrung out of tanned hides, multiple rinses of color wheels, and variances between 1988 and 1996 data.

Although the inputs and outputs of the system match within 3%, there are some abvious flows with this materials balance. The largest being the use of data from 1988 without assurance that past operating conditions are similar to current practices. In the ideal situation an unlimited project budget would allow weighing and sampling of all inputs, outputs, and in-process materials at each process step.

CASE STUDY 2: Leather Dyeing

The following is the first case study in this book to present actual financial data on the use of a specific material. Case studies following this one (see Chapter 7) will gradually become more complicated. For the first one, a simple accounting of how much the chemical costs to use will be performed.

Step 1: Review of Production Process

This small company operates a leather finishing operation and processes approximately 1700 sides of tanned leather, dyed leather, synthetic leather, and pigskin daily into finished product. Currently, all manufacturing operations are conducted in one structure with coloring, drying, and embossing or softening occurring in a linear progression through the building (see Figure 2.2).

The company finishes pretanned leathers through a series of spray and roll coating processes on conveyor belt lines. Sides of leather arrive at the facility already tanned, dried, and, in some cases, dyed or partially finished. Finishing operations consist of spraying coatings and colorings onto the leather sides as they travel along conveyorized lines. Electronic light beam detectors are used to start and stop the spraying machinery to minimize overspray. Chemicals are mixed in drums and containers at the sides of the process lines and siphoned to the spray guns. Following coating, the leather then passes through drying ovens and onto sizing, embossing, and quality control operations.

Glycol ethers are used as carriers in several of the water-based coatings and finishes applied to the leather. Glycol ethers also exist, to a limited extent, as mixtures with other regulated chemicals including acetone, toluene, and xylene. In accordance with state pollution prevention laws, the company must determine how much the use of this regulated chemical costs each year.

Step 2: Gathering Cost Data

We met with the plant manager and a line operator to discuss the use of glycol ethers. In our conversations, several interesting points were revealed. Cost data had to be extracted from purchase records and files on waste management. The company did not have a computer system to track materials, so information requests took longer to fulfill than in some other cases. Operator knowledge filled some data gaps. See Table 2.5.

1. Chemical purchase costs could not be determined within the time constraints of the project. The facility uses over 50 varieties of colorants, penetrants, dyes, carriers, and cleaners. Most of these have a blend of several solvents such as acetone, xylene and glycol ethers. The MSDS were not well enough organized to determine the components of each input. The company buys small quantities of these chemicals in 10- and

■ **FIGURE 2.2 Process Flow Diagram for Leather Coloring**

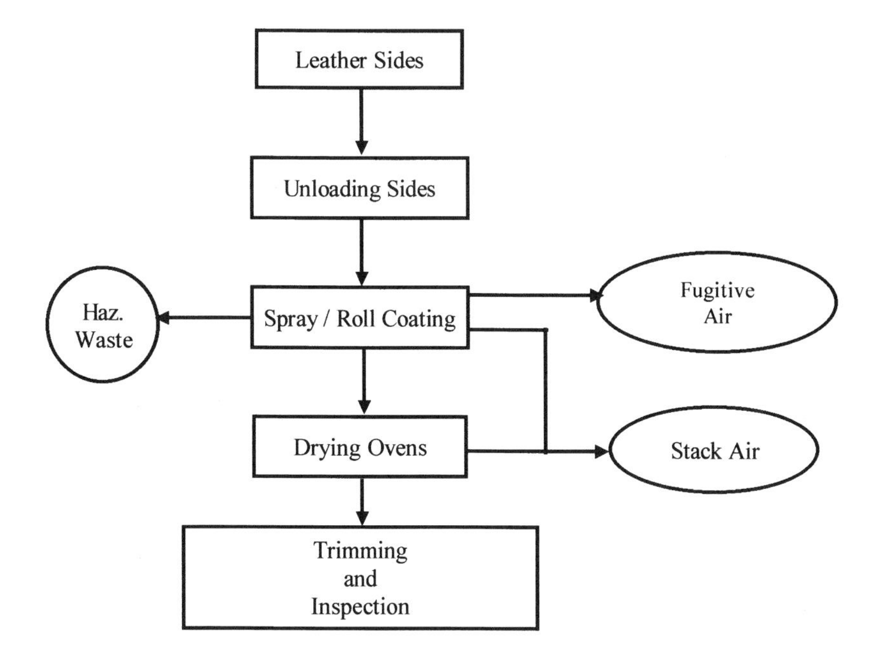

20-gal drums. All of these issues inhibited the development of a purchase cost for total glycol ether usage. Because of this, some cost categories were simply divided by four to show cost sharing between the 4 major reportable chemicals. Extra effort was then put into refining the unit prices and total number of units to compensate for this gross estimate.

2. The coatings are stored in drums in a separate room. Storage costs for these chemicals are considered minimal as there is no other competing use for the space. There may be an aesthetic cost or a safety hazard in having drums in use next to the coating machines, but no effort was made to quantify a connected cost.

3. Waste management costs include wastewater treatment (simple pH adjustment), testing for volatile organics in wastewater, and disposal of hazardous-waste paints. Treatment and testing costs were evenly divided among toluene, xylene, acetone, and glycol ethers. Waste disposal was allocated based on the fraction of glycol sent off-site compared to total hazardous waste. This fraction was determined by reviewing the available hazardous-waste manifests for the past year.

TABLE 2.5 Operating Cost of Using Glycol Ethers

Cost Item	Description	Cost Factor	Unit Price	Total Units	Total Cost
Purchase costs	Chemicals, glycol ethers	—	—	—	—
Storage/floor space	Ancillary chemical	N/A	—	—	—
	By-product (scrap)	N/A	—	—	—
	Emissions (waste)	N/A	—	—	—
Waste management	Treatment chemicals	25%	$33/day	250/year	$2,062.00
	Testing	25%	$2,400/year	1	$600.00
	Disposal	10%	$3,390/year	1	$339.00
Regulatory compliance	Safety/training equipment	25%	$120	1	$40.00
	Usage fees & taxes	100%	$1,100/year	1	$1,100.00
	Manifests, test reports	25%	$600/month	12	$1,800.00
Insurance		N/A	—	—	—
Production costs	$/unit of product	N/A	—	—	—
	Labor/unit product	25%	$20/day	250/yr	$1,250.00
	Other	N/A	—	—	—
Maintenance	Time	25%	$20/day	250/yr	$1,250.00
	Materials	—	—	—	—
Utilities	Water	25%	$2.36/100 ft^3	30,515 (100 ft^3)	$18,003.00
	Electricity	—	—	—	—
	Gas/steam	N/A	—	—	—
			Total annual operating cost		$26,444.00

4. Two of the four chemicals used in the coating process are common groundwater contaminants and pose a significant risk to the company in terms of contingent costs. These costs are not reflected in the insurance premiums paid by the company. The company occupies a new building having no history of pollution exposure events to date. Therefore all on-site liabilities are directly related to the use of these chemicals. Glycol ethers are not normally considered an environmentally persistent substance. Because of this no cost has been assigned to this category despite potential risks.
5. Production costs were determined through conversations with the plant manager and line operators. The coating lines require one person to supervise, mix, and refill the color drums. Changes to the process may effect the amount of leather colored per hour, the number of mixes required, or the degree of supervision. Therefore, this cost is valuable to know when alternatives are being considered.
6. The spray guns are cleaned after every color run. Changes of air filters, floor cleaning, and other maintenance connected with using the colorants are considered a part of this cost. Data on air filter costs or other materials used in process maintenance were not available at the time of the project. No cost has been included.
7. Mixing and some cleaning operations use water. This cost is probably only second to the cost of the chemicals themselves; however, this cost is unknown. In this case, the utilities are the largest cost, even without information on electrical use to move the conveyor systems or run the electron-beam curing table.

Unique Features of This Case

This case illustrates how little information can be available. Unfortunately, this tends to be more common with smaller companies. The lack of a computer tracking or inventory system also significantly reduces the availability of data. A company may save all the records required, but they may not be located on-site. If you are only visiting the site for a day, this means a wait for the information and potential communication difficulties about exactly what type of data required.

Sometimes the amount of data "available" is directly proportional to the disposition of the people you are interviewing. Extracting data from a file cabinet or computer requires effort, your contact may not be willing to expend the energy to assist you or not think your request is a priority. When you are under a tight timeline, consider the trade-offs of trying to get the information versus estimating it yourself.

In this case it was more effecient to leave purchase costs blank due to the large number of coatings containing glycol and the complete lack of information on their costs.

CHAPTER 3

THE BASICS OF MONEY

In order to fully understand and effectively use TCA, you must also understand the concepts behind the equations. Without overstating the obvious, calculating monetary values is at the core of the TCA, and, therefore, the basic concepts used in banking and accounting must also be understood. Among these concepts are the time-value of money, interest rates, inflation, depreciation, and income tax. The first three concepts will be dealt with in this chapter and the last two explained in Chapter 4. Although this material may seem too simple to consider, the subtleties of more complicated analyses rest on these very basic concepts, and so learning must start from these basics so that we can work our way up.

WHY USE THE TIME-VALUE OF MONEY?

There are as many ways to view a financial investment opportunity as there are ways to paint a house. For instance, you can easily paint half the house by standing on the ground, and more than half if you live in a ranch-style house. Or, with a little more effort, you can rent a ladder or scaffold. Experienced painters will tell you the ladder route allows you to perform a more thorough painting of the house. So it is with financial analysis. Although there may be methods to determine short-term (ranch-style) investments, larger projects having longer lifetimes require the use of financial analysis scaffolding. Extending the metaphor a bit more, it can be said that the financial scaffolding gives you something to support your conclusions, something others can grab onto as they look at your work and something that decreases the riskiness of making cost or savings projections.

THE USEFULNESS OF THE TIME-VALUE OF MONEY EQUATIONS IN TCA

There are only two major steps involved in conducting a TCA. First is data gathering and second is financial calculations. Data gathering consists of all activities and estimates necessary to generate baseline operating costs, and the projected capital and operating costs of the proposed project. TCA calculations require these numbers to complete the formulas and deliver a final value of the project's worth. Data gathering by itself has only immediate relevancy because these data say nothing of how the project will perform over time.

Out of the first part comes a need for a formula that will show project performance over time and what level of profit can be expected. There are many indicators of profitability: payback, ROI, NPV, and so forth. As will be explained in Chapters 4 and 6, each has strengths and weaknesses.

This second part must also present data to other decision makers in a format that will be readily understood and accepted. It may be possible to derive a unique formula that incorporates a high degree of accuracy, but if others cannot understand what the numbers mean, it will be more of a frustration than a benefit. There is also something to be said for not reinventing the wheel.

THE NEED TO USE THE CONCEPTS OF THE TIME-VALUE OF MONEY, PRESENT VALUE, AND INTEREST

TCA must acknowledge the time-value of money and impacts of interest rates and inflation on the potential investment. This is unavoidable for three major reasons:

1. The national and world economies function on the concept of time and growth. Through growth of the economy over time, the value of currency tends to decline (see the next section for detailed discussion of the time-value of money). TCA must therefore incorporate this concept of decreasing monetary value because most of the pollution prevention projects analyzed will have lifetimes of 5 to 15 years. Over this length of time, significant devaluation can occur (i.e., a dollar becomes worth only $0.64 when the inflation rate is a constant 3% over a 15-year period). This devaluation creates the potential for initial profit projections appearing overstated. For example, if an aircraft painting project is estimated to save $30,000 per year, the lifetime savings of the project could be erroneously estimated to be

$$\$30{,}000/\text{year} \times 15 \text{ years} = \$450{,}000$$

However, the impacts of a 3% inflation rate over the same period would reduce the actual value of this savings to $358,137, a difference of over $91,000. In this case, the cash flows from years 1 to 15 are worth only 79% of the estimated value. Refer to Chapter 5 for an explanation of how these numbers were calculated.

2. For an environmental manager attempting to "sell" corporate management on a pollution prevention project, the final numbers must be comparable and competitive with other capital purchases. This forces the use of terminology and methods that are already in use and accepted. The time-value of money and present value analysis is taught as a component of all MBA curricula and used to calculate loan payments, bond and stock yields, and most other types of financial forecasting. Serious analysis of financial investments always uses a time-value of money method to improve the accuracy of the projection.

3. For those unfamiliar with financial analysis, present value shows the effects of the savings over time. Although the proverbial "low-hanging fruit" (e.g., the easy projects) have payback periods well within acceptable ranges (1 to 3 years), the more complicated projects or larger investments may take longer to pay back. The people just starting out in pollution prevention may outrightly reject a longer-term project without knowing that a more thorough analysis would reveal the long-term savings (i.e., beyond 3 years). Present value captures these long-term savings and delivers a single value that can be compared against other competing projects.

WHAT IS THE TIME-VALUE OF MONEY?

An old adage oft-heard in manufacturing is "Time is money." More appropriate might be "Time changes the value of money." If a person were given the choice to receive $5 today or $5 next week, very few people would not choose the immediate money. Why? Another old saying offers the answer: "A bird in hand is worth two in the bush." People intuitively understand the concept even if it has never been explained to them. The reasoning behind the saying is that money in hand offers a choice about how it may be used. It can be spent right away or saved until next week. Having to wait until next week for the $5 means you may miss the opportunity to purchase an item, pay for food, or invest and make more money. This concept is called the *opportunity cost* of money.

A second factor in the value of money concerns the external growth of the national and world economies. Due to ever-increasing demand for resources and goods, the markets of the world produce more and more products. This growth is seen as a positive trait for all economies. But the growth over time also tends to make costs of goods increase as well. Everyone has experienced this when their grandfather tells them of how gasoline used to be a nickel a gallon or that hamburgers used to be a quarter instead of $1.59. Looking at this phenomenon from the dollar's perspective, the dollar is no longer capable of purchasing the same amount of goods as it did 10, 20, or 30 years ago. This results in a devalued dollar relative to the goods it can purchase. The rate at which this devaluation occurs is usually referred to as the *inflation rate.*

The impact of inflation on future profits must be accounted for to increase the accuracy of any financial analysis. The lost opportunity to invest the money at some time in the future must also be taken into account. For these two reasons, TCA uses a time-value of money method such as the present value.

The present value of a good accounts for all savings or costs accrued over its lifetime, and discounts these savings or costs by the interest rate to account for lost opportunity, inflation, and risk. Additionally, the present value interest rate can be chosen to represent a desired return rate. This rate might reflect a future investment opportunity's expected return on the investment, such as 8%. By incorporating interest into the calculations, we meet the need of devaluing money over time and the lost opportunity to invest the money elsewhere.

TIME-VALUE OF MONEY IN ACTION: NO PURCHASE NECESSARY

To illustrate the usefulness of the concepts of time-value of money, look at this example. Everyone in the United States seems to get sweepstakes entry forms in the mail with little stickers and a smiling face of some famous celebrity. The celebrity claims that average people such as yourself win every day; no purchase is necessary; just return the forms by the deadline. Suppose you won the sweepstakes! Say that you have just won the marketeer's sweepstakes for $10 million. The celebrity is knocking on your door and asking you to choose a prize package to award you. Here are your choices:

Option A: $10 million megaprize funded by U.S. Savings Bonds delivered at once and paying $10 million if held to maturity. The prize to be payed at $250,000 per year for 30 years, plus a final payment of $2.5 million on the 30th year. *We* would become instant multimillionaires, right?

Option B: $10 million megaprize funded by U.S. Savings Bonds, with $1 million cash at once, and $200,000 yearly, plus a $3.4 million payment in the 30th year. This option may not seem as good as the first, but *we* could still go really crazy with a million dollars on the first year.

Option C: $10 million megaprize funded by U.S. Savings Bonds, with $500,000 cash at once, $250,000 a year thereafter, plus a final payment of $2.5 million in the 30th year. Although not as appealing as instantly having a million dollars, this prize does have a higher annual payment than option B.

So which will it be? A? B? C? You cannot keep the celebrity host waiting too long or he will drive away and give your prize to someone else. There are two key issues to consider in this decision:

- The time-value of money
- The phrase "held to maturity" (a.k.a "always read the fine print")

First, look at the impact of time on your newly won millions. You know the value of money decreases as time goes on; thus, the worth of the annual payments and the end payment will not be as much as is stated.

Second, the phrase "held to maturity" means that the money is being derived from immature bonds. In this case, say it means the bonds will mature 5 to 7 years from the date of the award. If you know that your bank pays 3% per year on savings accounts, you can use this as a conservative interest rate, as a higher rate would erode the value of the winnings much more quickly. Because you have not learned how to work the present value formula

$$P = \sum \left[F \times \frac{1}{(1 + i)^n} \right]$$

yet, you call your accountant and she figures the current dollar value of the cash flows for each option. (Of course, she will want a significant cut of the prize money.)

Cash Flow A

Stated Value	*Year*	*Present Value*
$0	1–6	$ 0
$250,000/year	7–37	$4,187,516
$2.5 million	38	$ 813,066
Total value of A		$5,000,582

With option A, you must wait 6 years for the bonds to mature before you can draw money from them even though $10 million worth of bonds are given to you on the day you win. After the sixth year, you get annual payments ranging in present dollar value from $203,273 in the seventh year to $83,745 in the 37th year. In the 38th year, you finally get the last payment, which has been devalued to less than half its original amount! You actually get just over $5 million with this option.

Cash Flow B

Stated Value	*Year*	*Present Value*
$1,000,000	Now	$1,000,000
$200,000/year	2–29	$3,530,000
$3,400,000	30	$1,400,000
Total Value of B		$5,930,000

In this option, you at least get some cash up front ($1 million). Your annual revenues will range from $188,519 in the second year to $84,869 in the 29th year. The final payment is reduced to almost one-third in value, bringing the total present value of the prize to just under $6 million.

Cash Flow C

Stated Value	*Year*	*Present Value*
$500,000	Now	$ 500,000
$250,000/year	2–29	$4,547,104
$2,500,000	30	$1,029,967
Total Value of C		$6,077,701

The final option provides a modest income over time with $500,000 up front and a range of annual payments: $235,648 in the second year down to $106,086 in the 29th year. With the final payment, the total value is the greatest of the three options but nowhere near $10 million.

Of course, being outraged that you will not receive the actual value of $10 million, you tell the celebrity to take a hike and find some other sap. This example demonstrates how large sums of money ($10 million) lose significant value over large periods of time (30 years).

THE USUAL FINANCIAL ANALYSIS

Typically, the need for time-value of money calculations does not arise with the arrival of celebrities and sweepstakes checks. Instead, it is far more common to need to know whether purchasing an item is a wise use of money. Yet, surprisingly enough, most people, and businesses, do not incorporate the time-value of money into their financial decision making. Most managers in small to medium-sized companies use payback analysis to check the worth of a project. In brief, a payback time is calculated by dividing the expected annual savings into the capital cost of the project. For example, a $10,000 project that saves $3,000 per year would have a 3.3-year payback. The method is simple and can sometimes even be done without a calculator, no doubt contributing to its widespread use. However, there are serious drawbacks to this method that actually shortchange the user from potentially significant savings. The following examples demonstrate the dangers and lost opportunities incurred when using payback.

DISADVANTAGES OF PAYBACK CALCULATIONS

Suppose you must make a decision to invest in one of two new bioengineering technologies. You know very little about them except that each has an identical initial cost ($1 million) and each is guaranteed to have a 3-year payback. How can you make a proper decision without more information? What is needed is a picture of their cash flows over time. The cash flows for these two options are as follows:

	Technology A	***Technology B***
Initial investment	$1,000,000	$1,000,000
Year 1 revenue	$ 0	$ 333,333
Year 2 revenue	$ 0	$ 333,333
Year 3 revenue	$ 999,999	$ 333,333
Payback period	3 years	3 years

With technology A, you would not see a penny of your money until 3 years have passed. And as we demonstrated in the sweepstakes example, that amount is not really worth $999,999. With technology B, you get an annual payment of one-third of the original investment. This actually pays more in present dollars, and allows you to reinvest the money somewhere else (such as in solar-power cells).

Another situation could occur that payback analysis fails to reveal. Payback analysis is blind to the cash flows that continue to occur past the point where the initial investment has been recouped. Let us continue with the previous example of bioengineering technology.

Year	*Technology A*	*Technology B*
1	$ 0	$333,333
2	$ 0	$333,333
3	$999,999	$333,333
4	$ 0	$333,333
5	$ 0	$333,333

The cash flow diagram shown above exposes technology A as a one-shot deal. Three years comes by and your money is returned but no other additional income. Technology B continues to pay an annuity into the fourth and fifth years. If you evaluated these options using only payback analysis, you would have missed additional investment freedom, and additional cash flow. You can construct a cash flow diagram every time you evaluate a project, but that can become tedious. The more efficient way to quantify the dollars over time is to use a method such as (net) present value (NPV).

CHAPTER 4

THE MECHANICS OF THE PRESENT VALUE

This chapter covers the information necessary to perform present value calculations. It is divided into five sections to create a learning progression that parallels the steps of calculating the present value. The first section describes the terminology used throughout the book and references other chapters that provide greater detail on particular topics such as the use of comparison tools like the equivalent annual annuity. The second section helps develop a better understanding of how the present value calculation works, how to derive it, and the use of interest rate tables. The third section describes the differences between the cash flow diagram approach and the annuity approach for calculating the present value. Benefits and weaknesses of each approach are detailed. The fourth section explains depreciation, why it might be included in the TCA, how to figure depreciation, and how to determine the net present value (NPV). The chapter concludes by presenting alternative uses for TCA and the subtleties of interest rates, outdated prices, and tax liability.

TERMS AND DEFINITIONS

The terms and definitions that follow are used throughout this book. All have been taken from commonly used business and accounting vocabularies to minimize the creation of new acronyms and potential for confusion. The terms used herein are briefly explained when they first appear.

Capital Costs

Capital costs are all costs considered a part of the implementation of a new project or system. These include the research and design of the new system, obtaining necessary permits, changes to the production process or building, purchase of equipment, and disposal of the old process. It is recognized that research and design costs are not always allowable as expenses when reporting to the Internal Revenue Service. Keep in mind that Total Cost Assessment is not used for reporting to the IRS; it is used to make an informed business decision about a capital investment.

Discount Rate

A company's discount rate (also called the hurdle rate) is the interest rate used in evaluating projects. It reflects the cost of borrowing money to finance the project as well as a subjective, perceived level of risk involved. This rate is a confidential number and varies from company to company. It may even change from project to project, depending on market economics, investors' demands for returns on their stock, cost of debt and equity, ratios of debt and equity to value and the tax rate. For example, an environmentally aware and progressive company may allow a lower discount rate to be used for projects known to have a positive environmental impact, but use a higher rate for less "eco-friendly" projects.

The exercises in this book will always state the discount rate to be used. Your own analyses should always state the chosen rate in the list of assumptions. This allows others to compare the TCA with other financial analyses.

Operating Cash Flow

Operating cash flow refers to all regularly (annual) occurring costs, or savings, associated with the operation and maintenance of a system. These include both the conventional and hidden costs, such as chemical purchases; waste management; all fees and labor connected to regulatory compliance; annual permitting or testing; training; insurance costs; system maintenance; production costs; down time; storage and warehouse costs; water, natural gas, or steam; and electricity usage.

Remember that these costs must occur on an annual basis, or must be prorated to provide an annual cost figure. For example, a facility is required to develop a pollution prevention plan because it uses a specific chemical. This plan must be updated every 3 years. The cost for updating this plan must be included in the TCA prorated as an annual expense (i.e., one-third of the plan update cost per year).

Incremental Cash Flow

This term refers to the difference in operating costs between an old process and the new. For example, an old painting line costs $10,000 per year to operate. A new electrostatic paint line will cost only $4,000 per year to operate. The incremental cash flow is $10,000 (old system) − $4,000 (new system) = $6,000.

Interest Expense

A company may choose among several methods to fund a project. It may take reserve funds from within the company, it may obtain a bank loan, or seek an investor or venture capitalist. In the first case, there is a lost opportunity to invest the capital in a project of equal or higher return. This lost opportunity is accounted for when the present value equation is used (see what follows). However, in the second instance, the bank and investors will charge interest on the loan, which must be included as one cost of the project.

It has been our experience that banks are reluctant to finance environmental projects, even pollution prevention projects that have fantastic NPVs and eliminate or reduce hazardous wastes and chemical liabilities. The reasons for the lack of funding are numerous and many times client- or bank-specific. Some state and federal environmental agencies (Massachusetts, Illinois, and the U.S. Environmental Protection Agency's Office of Pollution Prevention and Toxics) have begun educating financial institutions about the benefits of pollution prevention. It is hoped that the future will not reflect the past in this arena. For most simple projects with small capital costs, interest expense is not a concern. Larger projects requiring outside financing should be analyzed using a more detailed TCA (see Chapter 5) that includes interest expense.

Opportunity Cost

Internal financing of a pollution prevention project means a lost opportunity to invest those funds elsewhere. Logically, this lost opportunity represents an expense to be counted against the project. However, the opportunity is embodied within the many financial analysis tools such as net present value. Comparing projects using these tools automatically considers the opportunity cost and no further treatment is necessary. Of course, the circumstance may occur when a pollution prevention project is evaluated against the investment of the same capital in mutual funds or stock options. In this case, factors such as the value of reinvesting capital in the physical plant and decreases in long-term liability also must be considered.

Depreciation

Depreciation is the process of allocating the purchase cost of a machine across its entire lifetime to represent the loss of value as a result of using the machine. A machine would lose its value because new technology makes it obsolete (the buggy whip is the classic example), newer models of the same technology are faster or have more capacity (upgrades in computers), and the machine's parts wear out or are used up (the engine of an automobile).

The machine may have some value at the end of its life. This is often called the *salvage value.* For example, a large metal degreasing tank purchased 30 years ago has long since been fully depreciated, yet it could still be sold for scrap metal and would have a salvage value. Salvage value should be included in a thorough Total Cost Assessment, but not all equipment has a salvage value.

How to show the loss of value of an asset has been a subject of debate for many years. The Internal Revenue Service has specified categories for all depreciable machinery to create consistency among taxpayers. Depreciation methods are discussed later in this chapter. This book uses straight-line depreciation for all options.

Depreciation is only relevant to TCA if you will be using the more detailed after-tax cash flow analysis. Counting depreciation as part of a before-tax analysis will inflate the results and frustrate your corporate financial managers.

Taxable Income, Income Tax, After-Tax Cash Flow

These terms are used to calculate the effect of income tax on the savings resulting from the pollution prevention projects. All corporations pay taxes just as all citizens of the United States are required to pay taxes. A project that saves a corporation money may impact the amount of tax that company pays each year. The impact could come through reduced operating expenses, which are a deductible expense for the company. This reduction in cost increases the before-tax cash flow, creating a larger number on which to base the tax rate, and more tax being taken out. In a very large corporation, the impact of one pollution prevention project may not have a significant impact, but for smaller firms, the differences are more noticeable. Therefore, the capital purchase cost of the project's new equipment would be most likely depreciated with the rest of the company's assets to reduce before-tax cash flow by as much as possible. If this is done, one *must* then use an after-tax cash flow analysis to account for these tax savings and the tax shield effect granted by the depreciated equipment. The following terms would then be incorporated into the TCA.

Taxable Income Taxable income is the operating cash flow minus the annual depreciation of the new project's equipment. Subtracting the depreciation decreases the total amount of income subject to taxation, correspondingly reducing tax payments.

Income Tax Rate The income tax rate is the percentage of taxable income paid by corporations. This rate varies as much for corporations as it does for individuals. An approximate rate can be used or the actual rate can be found by asking the corporation's chief financial officer or tax accountant. A company may not wish to disclose its tax rate, in which case an average rate can be used, provided the necessary caveats are added to explain how the average rate increases or decreases the financial feasibility of the project. This book uses a standard tax rate of 40%.

After-Tax Cash Flow After-tax cash Flow refers to the amount of profit remaining after federal taxes have been taken out. This number is sometimes called ATCF.

Annuity

An annuity is a single payment that remains the same for every year of the analysis. In TCA, there are several ways to calculate the net present value, one of which uses assumptions implied by annuities. For example, the costs of waste disposal for the first year might be assumed to be the same for all following years in the project's lifetime. The cost is then considered as an annuity, allowing compound interest tables for annuities to be used in determining the present value rather than calculating each year's present value separately. This method is discussed in more detail in Chapter 5. The main benefit to using annuities is the time savings; the main weakness is the assumption that costs will remain constant for the project's entire life. Annuities are also used in the comparison tool called equivalent annual annuity, which is explained in detail in Chapter 6, and summarized in what follows.

Net Present Value

Present value is the basis of TCA calculations and implies that the worth of an investment changes over time. The formula used to determine present value accounts for the effects of time on an investment opportunity. Present values in this book were calculated using the interest tables found in Appendix A. The concepts of time-value of money and present value are discussed in Chapters 3 and 4, respectively. *Net* in front of the present value is meant to show that the net present value is derived from the present value by subtracting the capital cost of the project from the present value. The final value of the project is usually referred to as the net present value.

Equivalent Annual Annuity

Differences between two projects may require the use of comparison tools other than the present value or net present value. When two options have different lifetimes, the net present values cannot be compared accurately. The net present values are then normalized by dividing each by their respective compound interest factors. See Chapter 6 for more detail.

UNDERSTANDING PRESENT VALUE

The formulas for present value are some of the most often used financial equations in the United States. The reasons for this and net present value's benefits over conventional payback analysis have been discussed. This section will explore the mathematics of present value equations and derive the equation that will be used in this book.

Historical Perspective

The concepts of the time-value of money go far back into history when commerce was conducted within the bounds of a kingdom, village, or tribe, and there was limited trade among these entities. Surpluses of goods existed or people found adequate substitutes for the goods desired. Improvements in agricultural techniques and mechanization allowed local economies to grow fast enough to create large demands for goods and services. This demand produced a scarcity of goods. The specialization of goods decreased the number of suitable substitutes for a desired good. People could either wait for supply to increase or pay more for the same good. Those who could pay more did not have to wait. This resulted in price increases for the goods in short supply. As the prices began a trend of increases, people expected this trend to continue, thus further increasing prices. These price increases were collectively named inflation. As prices increased for all goods, the purchasing power of the currency decreased. As grandparents are fond of reminding us, a dollar does not go as far today as it did back in 1927. As time progressed, trade with other continents increased and introduced a third major factor influencing the time-value of money: exchange rates. Fluctuations in currency values between two countries also contribute to inflation as the amount of trade between them increases. A drop in the value of the U.S. dollar compared to the Japanese yen makes importing products from Japan more expensive. The prices of these goods therefore would increase when brought to market.

The banking industry also contributed to the time/money issue by charging interest on funds loaned and paying interest on deposits. Charging interest on a loan dates back almost as far as the use of currency itself. Interest payments on loans were considered the cost of borrowing money. Initially, banks in the United States were holding places for money that were considered safer than keeping it in mason jars in the back yard.

However, if the money was in the bank vault, then the depositor could not use it to invest in another venture. Therefore, the opportunity to earn money was lost by leaving the money in the bank, unless the bank paid the depositor a return on the "investment." This lost flexibility to invest the money at any given moment also reinforced the time-value of money theory by creating what is known as opportunity cost.

It was not until after World War II that the concept of present value was applied to corporate financial decision making. In 1959, Modigliani and Miller[1] provided a more mathematical formulation and unity to the subject of corporate finance theory. The clarity of this "economic science" was built around the paradigm of equilibrium in perfect markets (discussed further in Chapter 11). Suffice to say, all corporate finance theory prior to the late 1950s soon became the subject of ridicule and dismissed as riddled with inconsistencies, anecdotal evidence, and motivated by the subjective real-life experiences of its proponents. In many ways, it resembled the current state of theories in the

[1]F. Modigliani, and M. H. Miller, "The Cost of Capital, Corporate Finance, and the Theory of Investment," *American Economic Review*, Vol. 48, No. 3, pp. 261–297.

field of physics. Many theories abound, yet there is no single unifying theory. Pre-1950s corporate finance was said to be equally divided between pursuit of corporate growth and the humanitarian aspects of business.

In today's business world, not considering the time-value of money in business decisions automatically places the project and investor at a disadvantage. To say that the time-value of money is unimportant results in the loss of value in the investment simply because all other investors, banks, economists, stockbrokers, and business persons think it is important. If a manufacturing company approaches a bank for a loan to purchase new process equipment, that bank will evaluate the risk of the proposal using the time-value of money and charge interest on the loan. Without considering the effect of loan interest payments, the receiver of the loan will pay more back to the bank than originally anticipated. The "extra" capital required to pay back the loan will decrease the original profit margin, making the project look like a poor investment. In reality, the initial projections were overly optimistic because they did not include the time-value of money.

Simple Interest

The first step in working with present value calculations is to understand the formulas for simple interest, compound interest, and how they are derived. The following symbols are used throughout this chapter:

I = annual interest rate
n = number of years
P = present sum or current dollar amount
F = future sum, made n years from now

Simple interest is a fixed percentage of a dollar amount given at the end of a period of time. It is the root from which more complicated formulas evolve. An example of simple interest is a 10-year loan of $100 at 10%. At the end of 10 years, the amount to be repaid is $110. Simple interest may be used for a loan between friends and family members, but is never used in business. The formula for simple interest is

$$F = P + P \times I \qquad \text{or more simply} \qquad F = P(1 + I)$$

Compound Interest

Compound interest entails adding the interest for one period to the present sum and using this new sum to calculate the interest for the next period.

By borrowing from the simple interest example described earlier, an amount ($100) is deposited in a savings account at interest rate I (10%) for n years (5 years). The calculations to arrive at the total interest earned at the end of five years are as follows:

Year	*Compound Amount*		*Simplified Equation*	*Value*
0	P		P	$100
1	$P + Pi$	or	$P(1 + I)$	$110
2	$P(1 + I) + P(1 + I)I$	or	$P(1 + I)^2$	$121
3	$P(1 + I)^2 + P(1 + I)^2 i$	or	$P(1 + \mathrm{I})^3$	$133
4			$P(1 + I)^4$	$146
5			$P(1 + I)^5$	$160

This equation can be generalized by factoring out the $(1 + I)$, as shown in the third column. This results in the standard compound interest formula:

$$P(1 + I)^n$$

The preceding table illustrates just 5 years of payments made at a 10% interest rate. If the loan described in the simple interest example were made using compound interest, the total amount due at the end of 10 years would be $259 instead of $110. In 20 years, the amount has grown to $672, and in 100 years, the $100 amount would have grown to $137,861! This is the fundamental principle (pun intended) behind all investments and retirement funds.

Present Value

The compound interest equation can calculate the worth of a current dollar amount in the future, and it can also be reversed to yield the current worth of a future amount. This variation of the formula is called the present worth interest equation:

$$P = F \times \frac{1}{(1 + i)^n}$$

Notice that it is simply a reciprocal of the compound interest equation. When the new equation is applied to the preceding example, it shows the value of the $100 through time in current or present value. Typically, this action is called inflation, and an inflation rate is used when in the macroeconomic setting. In financial analysis of projects, a slightly different rate, called the discount rate or hurdle rate, is used instead of the inflation rate. An analysis of the cash flows over time for a project is called a *discounted cash flow analysis* because the flows of money are discounted, or reduced, by applying the chosen hurdle rate.

A company's hurdle rate is established by corporate management to sufficiently compensate the company for the risks taken when investing in the project. This rate may incorporate the interest rate (cost of capital) charged by a bank for a loan to purchase equipment or the cost of using corporate reserves if the purchase is internally financed. The hurdle rate is unique to each project based on the certainty of the expected outcome. A high-risk venture such as mining for ore would have a higher hurdle rate than investing the same amount in U.S. Treasury Bonds. The following table illustrates the effects of discounting the previous $100 at 10% for 5 years.

Year	*With 10% Discount Rate*	*Without Discount Rate*
0	$100	$100
1	$91	$100
2	$83	$100
3	$75	$100
4	$68	$100
5	$62	$100

In this case, after 10 years, the $100 investment is worth only $39 in current dollars; in 20 years, its worth drops to $15; and in 100 years, it is worth less than $0.01!

The Effects of Interest Rates

In the preceding example, a discount rate of 10% is used to show the loss of value over time. This process can be accelerated by choosing a higher rate. At a 20% rate, for example, $100 is worth only $40 after 5 years, and drops to less than $0.01 in just 40 years. The opposite is true for decreasing the hurdle rate. A 5% rate would yield $78 in 5 years and take 189 years to lose value to $0.01.

Understanding the effects of interest rates on investment value are extremely important! Knowing what will happen to the value of your investment if the rate is changed will help you answer questions raised during the review of your project proposal by management, a bank loan officer, or an interested investor. Do not proceed further until these concepts are clearly understood.

Varying the hurdle (discount) rates on pollution prevention projects might look like the following:

- A higher hurdle rate for an unproven technology such as a new chemical for cleaning parts. This investment may cause product quality problems if the parts are not properly cleaned or there may be unknown acute or chronic worker exposure risks.
- A lower hurdle rate for a low-risk project even if it requires a large capital investment such as modernizing machine tool centers or segregating liquid-waste streams for potential reuse.

Interest Rate Tables

Many calculators can calculate present values with the press of a button. If you do not have access to such a calculator, you can use compound interest tables. These are tables of numbers showing a factor that can be conveniently multiplied by either the present sum or future sum.

Follow along with this example to learn how to use these tables. A life insurance salesperson can give you a $160,000 policy, payable on your death, if you live 20 more years and pay a onetime membership fee of $10,000 up front. Being savvy in the ways of the time-value of money, you quickly pull your present value tables from your back pocket.

First, you must determine the lifetime of the project in years. Because you must live another 20 years before the policy matures, this will be the lifetime and the number of periods.

Next, choose a hurdle rate based on the relative risk of the project and expected outcome. Investing $10,000 up front seems risky, so you might choose a higher interest rate, say, 15%.

If you had this book handy, you could turn to the interest tables in Appendix A. Table 4.1 shows an abbreviated version of the same tables and highlights which columns and rows to look at.

Look for the appropriate number of periods on the far left column and the desired hurdle rate on the top row. Follow the column beneath the hurdle rate and row next to the period until they intersect at one of the listed decimals. This is the present value factor for the given period and hurdle rate.

■ TABLE 4.1 Factors to Change a Single Future Value into a Single Present Value

	Interest Rates				
Period	**1%**	**3%**	**8%**	**10%**	**15%**
1	0.9901	0.9709	0.9259	0.9091	0.8696
2	0.9803	0.9426	0.8573	0.8265	0.7562
3	0.9706	0.9152	0.7938	0.7513	0.6575
4	0.961	0.8885	0.735	0.683	0.5718
5	0.9515	0.8626	0.6806	0.6209	0.4972
6	0.9421	0.8375	0.6302	0.5645	0.4323
7	0.9327	0.8131	0.5835	0.5132	0.3759
8	0.9235	0.7894	0.5403	0.4665	0.3269
9	0.9143	0.7664	0.5003	0.4241	0.2843
10	0.9053	0.7441	0.4632	0.3856	0.2472
11	0.8963	0.7224	0.4289	0.3505	0.215
12	0.8875	0.7014	0.3971	0.3186	0.1869
14	0.87	0.6611	0.3405	0.2633	0.1413
16	0.8528	0.6232	0.2919	0.2176	0.1069
18	0.836	0.5874	0.2503	0.1799	0.0808
20	0.8196	0.5537	0.2146	0.1487	0.0611

Multiplying this factor by the known future amount will yield the value of that amount in current dollars. So that,

$$\text{Present value} = \$160{,}000 \times 0.0611 = \$9{,}779$$

The insurance salesperson becomes a little uncomfortable when you show that your $10,000 investment will yield only $9,779 after 20 years. The salesperson makes an excuse, leaves, and is never seen again.

To summarize:

To Find	***Multiply This Amount***	***By This Factor***
Present amount	Future amount	Present value factor (always < 1.0)
Future amount	Present amount	Compound interest factor (always > 1.0)

A Note on Accuracy

The present value factor and the compound interest factor are mathematical reciprocals of each other ($x = 1/y$ and $y = 1/x$) except as noted in the following. Most compound interest/present value tables are calculated to only four decimal places. This is accurate enough for most applications, especially in banking where loans rarely exceed 30 years. However the n values in these tables often extend down to 100 periods or more. At higher interest rates, such as 10%, certain rounding errors are introduced as n increases and the space allotted to print the decimal places stays fixed. This can be confirmed by multiplying a principal amount by the compound interest factor for the highest value of n in the table. Then multiply this future amount (at the nth period) by the present value factor listed in a corresponding present value table. The value of the resulting future amount in current dollars should not exceed the initial principal amount. The degree to which this number differs from the original principal amount reflects the rounding errors of the table. This fact is mentioned in the event environmental managers/engineers should desire to calculate costs or benefits of projects with long lifetimes (greater than 30 years) such as major highway expansions, hydroelectric or nuclear facility construction, and natural resource conservation programs. Because the costs of these projects tend to be measured in hundreds of millions of dollars, the rounding errors could be more noticeable.

What about WACC?

Some companies may desire to use a more precise discount rate for the net present value calculations that incorporate risks from debt, interest on equity, and the firm's debt-to-equity ratio. One alternative commonly used is the weighted average cost of capital (WACC).[2] This is the finance term for an average cost of capital derived from several other costs, ratios, and interest rates. WACC is composed of the cost of equity capital (roughly the percentage return on publicly traded stock of the company, i_e), the cost of debt (the interest rates charged by the bank on outstanding loans, i_d), the market values of total debt, (D) and equity (E) of the firm, and the value of the firm (equity minus debt $V = E - D$), and the corporate tax rate (t). For those

[2]Richard Torberg, "Capital Budgeting for Environmental Professionals," *Pollution Prevention Review*, Autumn 1994, p. 459.

people wanting to calculate WACC and use it, the following equation is provided:

$$\text{WACC} = i_e(E/V) + i_d(1 - t)(D/V)$$

CALCULATING THE PRESENT VALUE

Cash Flow Analysis

Business schools teach their students to draw cash flow diagrams for all incoming and outgoing capital. These diagrams show total receipts and disbursements for each period for the life of the project. This book will not go into the development of cash flow diagrams because they are not used in TCA. The reader is refered to a reference book on finance or accounting, such as *The Ten Day MBA* by Steven Silbiger, for additional detail. Once a cash flow diagram or bar graph has been completed, the present value equation is applied to each cash flow to convert it into current dollars. This highlights any unusual features of the investment such as a greater number of dispersements in the beginning than at the end of the project, or large or disproportionate receipts at any time during the project. Next, all the individual present values are totaled, resulting in the present value of the investment. This approach shows details of all capital involved with the investment as time progresses. MBA programs stress this method of cash flow analysis to build competence in handling numbers and the time-value of money concept. Despite its benefits to helping build understanding, it is a time-consuming approach when applied to projects in the private sector. Imagine having to draw diagrams, list the cash flows, calculate the present values of each, and then total them for five, ten, or fifteen alternatives being considered, each having a 10-year series of dispersements. That amounts to 155 to 460 operations required to compare the alternatives!

The compound interest tables can again provide a shortcut in determining present values. When dealing with equal annual payments, the present value of each payment can be summed and expressed as a new equation:

$$P = \sum \left[F \times \frac{1}{(1 + i)^n} \right]$$

The Greek letter sigma means "perform this calculation for the desired number of periods from 1 to n, and add them together." Most financial/business calculators have this function built in, but if yours does not, the compound interest tables for annuities can be used. An annuity is an equal annual payment. The compound interest tables for annuities list numbers that incorporate the increased value derived from each successive annuity. The user simply locates the number of periods (n) and the interest rate (cost of capital) and finds the corresponding number. Multiplying this number with the present total annual cash flow will yield the present value of all future annuities. This eliminates having to perform numerous present value calculations and finding their totals. The simplicity of this approach can be appreciated if one is required to analyze five separate 10-year projects. Under the former method, this would require 50 individual calculations versus five using the annuity approach.

Annuities

Notice the subtle difference between using the cash flow diagram approach and using an annuity table. In the cash flow diagram approach, each cash flow is carefully documented no matter when it occurs or what it's size. The annuity approach is more generalized because it operates with the assumption that all cash flows are equal across years 1 to n. One must be careful to fully review the types of cash flows involved in each project. It is highly likely that future expenses will increase after some period between 1 and n or that revenues will increase as time progresses. If these expenses or revenues can be averaged over the life of the project, then the cash flow for any one year will be the same. The following example illustrates the process for averaging a cash flow to annuity. Suppose a company pays a contractor $10,000 once every 3 years to clean out sludge from a storage tank. Over a 10-year lifetime, this tank would be cleaned out three times ($30,000), and at the end of the tenth year, there would be some sludge left in the tank. If the cost for cleaning were equally distributed over 10 years, this would equal a $3,000 annuity. At 10%, the present values for both the cash flow diagram and the annuity method are detailed in what follows:

Time (Years)	*Cash Flow Diagram Method*	*Annuity Method*	*Time (Years)*	*Cash Flow Diagram Method*	*Annuity Method*
0	$0	$0	6	$5,645	$1,693
1	$0	$2,727	7	$0	$1,539
2	$0	$2,479	8	$0	$1,399
3	$7,513	$2,253	9	$4,241	$1,272
4	$0	$2,049	10	$0	$1,156
5	$0	$1,862	Total PV	$17,399	$18,429

The table shows a difference in current dollars between the two methods. Using the average annuity method overestimates the present value by 5%. How would this overestimate affect a TCA?

Most cost categories (i.e., chemical purchases, utilities, waste treatment, etc.) occur on an annual basis or are recorded as annual totals. Nonannual expenses (such as the tank cleanout just illustrated) are more often the exception than the rule. When averaged as an annuity, nonannual expenses will yield an overestimate of the present value cost. As long as you consistently apply either method, that is, you do not switch from one mode of calculating to another in the middle of an analysis, these final numbers will be comparable to others. The degree of inaccuracy introduced by using average annuity is no greater than the unpredictability of other factors influencing profit forecasting (the national economy, corporate spending policy, shareholder demands, etc.).

Annuity-based calculations trade time-efficiency for a marginal decrease in cost estimating accuracy. Given that the majority of operating expenses will occur on an annual basis, this marginal inaccuracy can be justified, or avoided in cases where the reader deems it necessary to perform the cash flow diagram approach. The examples in this book use the annuity approach for the main reasons that nonannual expenses are rare, and the time saved in calculation offsets the marginal inaccuracies introduced.

CALCULATING THE NET PRESENT VALUE

The Effects of Depreciation and Tax

According to the *American Heritage Dictionary*, the term depreciation means to "make or become less in price or value." The significance of this term can be appreciated by anyone who has owned a new car. After signing the loan, putting the keys into the ignition and driving off the automobile dealer's lot, the new car owner typically hears a subtle *wump!* as the car loses about 20% of its value. The car will continue to be worth less and less until, for a time, the cost to repair the car is more than the value of the car. After about 30 or so years though, the car increases in value because it earns the title of antique. Why does the car first lose value and then regain it? The car is considered a durable good and as such has a fixed life measured in years. Every year, more and more of the car's usefulness and ability to perform its function is used up. At some point, the car may not function. If sold for parts or scrap metal, the car would have *salvage value.* If by some chance the car significantly outlives its expected life, it becomes a rare object in short supply. As discussed in the beginning of this chapter, limited supply translates to increased cost or perceived value.

All the assets, equipment, buildings and materials a company owns are considered units of capital that could be sold at any time. One of the main reasons for tracking depreciation is to have a continuous monetary value of the company's assets. These capital investments in manufacturing equipment, pollution control technologies, and durable goods have a useful life. The depletion of this useful life can be attributed to four major causes:

1. ***Natural Deterioration:*** Breakdown, corrosion, or decomposition from exposure to wind, rain, soil, bacteria, and chemicals.
2. ***Physical Use:*** Use of the item that results in wear, shock, tear, etc.
3. ***Obsolescence:*** Development of new items that are more efficient in the same task, making the original unit uneconomical to operate.
4. ***Inadequacy:*** Demand rises to levels unexpected when the item was initially purchased, resulting in capacity shortfalls.

Some people credit this concept of obsolescence (older technologies and machinery for newer ones before their useful life is up) as a primary driving force in the development of the United States. For example, the sailing ship was replaced by the steamship, horses were replaced by streetcars, then buses, and now automobiles. A counterargument can be made using TCA that perhaps our society was too hasty to embrace new technology without fully understanding some of its consequences or costs (i.e., automobile pollution, nuclear waste, ozone depletion).

Another main purpose of depreciation is to decrease the amount of revenues subject to taxation by the government. In this way, depreciation functions as a "tax shield." The Internal Revenue Service allows companies to depreciate different types of assets, but sets limits on the amount and time period for depreciation. Over the years, accountants have developed many methods for depreciating assets, but as of 1981, the IRS introduced a standard to be used when preparing income taxes. The other forms are still used for internal accounting and project analysis. Four basic depreciation methods are described in what follows and the differences between each explained.

Straight-Line Method

This method has been around the longest and is the most simple to use and understand. The straight-line depreciation model assumes the value of an asset decreases at a constant rate (in a straight line). It divides the total capital cost of the asset by the expected life of the equipment. If, for example, an asset of $2,000 value had a 10-year lifetime, the depreciation would be $200 per year (assuming the asset had no salvage value). This method provides a quick number for use in additional calculations, but does not have the same tax benefits of the other methods. The exercises in this book employ this method because it is simple and the helps keep the focus on the TCA instead of calculating depreciation values.

Declining Balance Method

The declining balance method introduces a faster rate of depreciation in the early portion of the asset's life. A fixed percentage is multiplied by the book value of the asset at the beginning of each year. Using the same $2,000 asset but assuming a 30% depreciation rate yields the following depreciation amounts:

Year	*Calculation*	*Depreciation Amount*
1	$2,000 × 0.3	$600
2	$1,400 × 0.3	$420
3	$980 × 0.3	$294
4	$686 × 0.3	$206
5	$480 × 0.3	$144

The depreciation amount becomes smaller every year, but never completely depreciates the asset. If the asset had no salvage value, then adjustments would have to be made at the time of disposal of the asset to correct these differences. The double declining balance method is a variation often used for income tax purposes in which the maximum allowable depreciation rate is double the straight-line depreciation rate.

Declining Balance Switching to Straight-Line Depreciation

Under pre-1981 federal tax law, it was allowable to depreciate an asset over the early portion of its life using a declining balance and then switch to straight-line for the remainder of the asset's life. The switch usually occurs when the straight-line amount is greater than the declining balance amount for the next year. Using this method on a $20,000 asset yields these depreciation figures.

Year	*Calculation*	*Depreciation Amount*
1	$20,000 × 0.3	$6,000 (declining balance)
2	$14,000 × 0.3	$4,200 (declining balance)
3	$20,000 ÷ 5	$4,000 (straight line)
4	$20,000 ÷ 5	$4,000 (straight line)
5	$20,000 ÷ 5	$1,800 (balance of value)

At the end of the second year, the depreciation changes over to the straight-line basis because in the third year, the declining balance would yield $2,940, which is less than $4,000. The fifth year depletes the asset before the entire year is through. And, therefore, only $1,800 can be deducted.

Accelerated Cost Recovery System (ACRS)

The passage of the 1981 Economy Recovery Act established a uniform method of classifying and depreciating assets. The ACRS covers depreciable assets in service during and after 1981. There are only four classes of property (3, 5, 10, and 15 years), with depreciation rates prescribed by IRS tables for each year of each class. For example, the rates on a 3-year property are 25% the first year, 38% the second year, and 37% the final year. This method is an adaptation of the declining balance switching to straight line. ACRS assumes no salvage value for the property at the end of its lifetime. Any value recovered is subject to a tax liability. Using the $20,000 example from the previous method and substituting IRS-specified percentages yields these depreciation figures.

Year	*Calculation*	*Depreciation Amount*
1	$20,000 × 0.15	$3,000
2	$20,000 × 0.22	$4,400
3	$20,000 × 0.21	$4,200
4	$20,000 × 0.21	$4,200
5	$20,000 × 0.21	$4,200

In this example, the switch between the two methods occurs at the end of the second year. No value remains at the end of the fifth year.

This book makes use of an option offered under the ACRS program to allow straight-line depreciation at specified lifetimes. In the case of assets classified in the 5-year category mentioned earlier, the taxpayer can choose either a 5-, 12-, or 25-year depreciation period. If the asset were covered under the 10-year category mentioned earlier, the taxpayer can choose among a 10-, 25-, or 35-year depreciation period. Although it seems strange to assign a 12-year life to an asset in the 5-year category, there is a reasonable explanation. ACRS categories are accelerated depreciation schedules that allow businesses to write off their assets quickly, more rapidly than their actual useful lives. Automobiles are depreciated as 3-year property, but may last 10 years. For income tax purposes, it does not matter how long the car lasts, only that its capital cost be fully depreciated in 3 years. It is sufficiently accurate to use the straght-line method of depreciation, as long as a reasonable lifetime is chosen. To select a lifetime for an asset, refer to ACRS tables available from your company's tax accounting department. Most production equipment will fall into the 5-year category, and you must decide whether a 5-, 12-, or 25-year depreciation period is warranted.

Asset Lifetime and Tax Liability

ACRS tables will also provide a common basis between budgeting analysis performed to select the project and the tax accounting conducted to depreciate the equipment after it is purchased, installed, and operational.

Why would this be important? For smaller companies, the projections of cost savings are more critical to the long-term operation of the firm. Larger companies have the advantage of burying the profit/loss in the numbers of many production units and divisions. A small company may have only one production unit on which the entire company's financial performance is measured. Changes made to this production unit would be immediately reflected in the revenues of the company. Therefore, an overly optimistic projection of cost savings from a pollution prevention project could result in a shortfall of capital in the future. Using a short lifetime in the analysis of a project and a longer lifetime during the tax accounting process could create such a loss. The amount of depreciation directly affects the amount of tax paid. The larger the amount depreciated, the less tax paid, and vice versa. Whereas depreciation is not a cash flow per se, tax payments are. Choosing to depreciate an asset rapidly (by choosing a short lifetime) during the TCA will create a false image of decreased tax liability until it comes time to pay taxes.

DEPRECIATION IS NOT A CASH FLOW

Unlike rebates, taxes or dividends, depreciation is not a true cash flow in that it transfers no revenue to the company. Depreciation is used to reduce the taxable income of the corporation and in turn decreases the amount of taxes paid out. In this manner, it functions as a tax shield, but at no time is there ever a check received by the corporation for the amount of depreciation declared. This concept is sometimes difficult to grasp when actually performing the tax shield calculations. The following example demonstrates the tax shield concept. A company purchases a $7,000 piece of equipment that earns them $10,000 in gross revenues. It has a 7-year lifetime and is depreciated at $1,000 per year. The company never receives the $1,000 payment. The only real savings comes from the avoided income tax, as illustrated in the following table:

Operation	*With Tax Shield*	*Without Tax Shield*
Before-tax income	$10,000	$10,000
Subtract depreciation	($1,000)	$0
Taxable income	$9,000	$10,000
Subtract tax (30%)	($2,700)	($3,000)
Add depreciation	$1,000	$0
After-tax income	$7,300	$7,000
Tax shield amount	$300	$0

By depreciating the equipment, the company saved $300 in taxes. This process will be incorporated into TCA methodology and reviewed in greater detail in Chapter 7.

VARIATIONS AND LIMITATIONS OF THE PRESENT VALUE

As with any method, there are limits to what can be plugged into the equations to obtain reliable answers. This section describes the subtleties of the present value calculation and provides tips to keep from making credibility-gouging mistakes.

Present Value of Past Prices

The TCA approach takes place in the present and looks toward the future to anticipate savings or costs when completing a particular project.

Current market prices for materials and services are used to develop the capital and operating cost totals. Sometimes it is not possible to implement the project as soon as the analysis is completed. During this period of time, prices for services, materials, and utilities may have changed significantly since the analysis was performed.

If the analysis involves purchases not made in the current year, then the current market value of these materials must be used. The manager/engineer should determine the current price for each cost element. It would be a mistake to either assume prices have not changed or to attempt to increase past prices by a certain factor (Consumer Price Index, inflation rate, etc.). This latter approach may seem attractive, but the principle of price independence prohibits its use. Each commodity has a value determined by factors that are independent of the values of other commodities, aggregate indices, and overall inflation rates. Just because the price of corn rose $1.00/bushel does not necessarily mean the price of gasoline rose by the same amount or percentage. Because of this independence, no analogs between past and current prices can be drawn. The only reliable method for updating a TCA is to determine current market prices for all services, materials, and utilities mentioned in the older TCA.

For example, a division of a large corporation pilot tests a new technology on a process common to all divisions. The test requires more than 1 or 2 years to complete and issue the final report to the other divisions. At that point in the future, the costs of materials may be different than when the pilot test began. The price difference may be due to market demand, inflation, or government taxes. If the environmental managers do not use current prices in their analysis of whether the pilot technology is profitable for their divisions, the results may present overly optimistic cost savings.

A real-life example of unpredictable price increases centers around the regulation of chlorinated solvents. Prior to the global resolution of the Montreal Protocol of 1990, even though the demand for chlorofluorocarbons (CFCs) and other chlorinated solvents was strong, these chemicals were relatively inexpensive. Trichloroethylene's (TCE) going rate in 1989 was $3.95/gal and CFCs were roughly $8.90/gal. Following the Montreal Protocol, which established international phaseout guidelines for CFCs, market prices for these chemicals skyrocketed. In 1993, TCE sold for $12.80/gal (a 324% increase in 4 years). Trichloroethane (TCA) rose from $5.20/gal in 1991 to $16.71/gal in 1993 (a 320% increase in just 2 years). CFCs experienced an 800% increase in price from 1990 to 1993! Note that these price increases took place during years when the national economy was actually in recession. A review of the inflation rates, Consumer Price Index, or Producer Price Index would not have revealed this increase. During this period, roughly 40% of the post-Montreal Protocol market prices were federal taxes imposed to accelerate a chemical usage phaseout. Suppose that in 1992, an environmental manager tried to update a 1989 cost analysis for replacing chlorinated solvents. By using a standard 3% inflation rate instead of checking for the current market prices of the solvents, the new cost analysis would be terribly inaccurate. The TCA would not reveal the alarming trend in chemical purchase costs or the potential savings from switching to an nonchlorinated alternative solvent.

Present Value Falls Short on Resource Allocation

As will be discussed in Chapter 13, the values of both tangible (trees and minerals) and intangible (clean air and stable climate) natural resources are not adequately represented in this model of decreasing purchasing power over time. In the case of tangible natural resources, improvements to extractive technologies (i.e., better chains saws, more efficient mining drills) allow rapid depletion of remaining resources, while maintaining the illusion of steady supply of the resource through lowered prices. For intangible resources, markets cannot represent the total value of many life support processes and complex natural processes such as nitrogen fixation, global heat distribution, and climatic stability. These ecosystems are larger and more complex than economic systems, having impacts on multiple species, continents, and other processes that may be unknown. Already unforeseen consequences of ozone layer thinning have cost the public more than the combined taxes on chlorofluorocarbons (CFCs) generate.

This incongruity between resource valuation and the time-value of money will eventually need to be rectified. Unfortunately, the current economic model is the one the world accepts and uses. Until it changes, the prevailing methods must be followed or financial analysis projections must clearly outline the reasoning for not including the time-value of money. Given that a change in macroeconomic policies is not probable in the near future, we will proceed to discuss the mechanics of present value analysis.

Use of TCA for Budgeting Projects

Total Cost Assessment provides detailed insight into the costs of production processes and pollution prevention projects. The analysis is performed to narrow the choice among several competing alternatives or to determine the financial feasibility of a preselected technology. Many companies have strategic management plans or initiatives establishing numerical goals for cost reductions, waste minimization, or decreases in chemical usage. Good examples are participating in the U.S. Environmental Protection Agency's 33/50 program. This voluntary initiative solicited industries to reduce facilitywide usage of certain listed chemicals by 33% and 50%. Many industries participated in this program. To be sure, the managers and engineers of the participating companies had to analyze the processes using these chemicals to determine the feasibility of reducing usage. TCA can assist in planning projects to achieve such goals by narrowing the search for alternatives from the outset. TCA can provide not-to-exceed budgets that ensure desirable return on investment (ROI) or NPV. If a company requires a certain ROI for a project to be accepted by the corporate approval process, TCA can be used to back-calculate cost ranges to ensure the ROI is met. Current operating costs can be used to determine a base price for equipment purchase or a maximum allowable expenditure on chemical purchases, once the desired ROI or NPV is known.

The first step is to establish a numerical goal for reducing a waste stream, labor cost, chemical usage quantity, or other cost category. Next, the company's standard or desired ROI and corporate tax rate must be determined. Finally, the existing operating budget for the process being analyzed must have cost data for all the categories affected by the numerical goal.

The following example illustrates the methods for meeting expected ROIs. The Spark-O-Matic Corporation has joined the EPA's Design for Environment Printers Project. Desiring to be leader, the company president pledges to redesign the box printing process in his facility to reduce solvent usage by 80%. The company's environmental engineer is directed to make this project a priority but to ensure that all financial performance indicators are met. Spark-O-Matic requires a 15% ROI on all investments. Current operating costs on the box printing line average $130,000 per year, with most of this going to purchases of solvent-based inks, cleaning solutions, and hazardous-waste disposal. The engineer uses the president's desired decrease in solvent usage (80%), the company's desired ROI (15%), the corporate tax rate (40%), and the annual box printing process operating budget ($130,000) to develop her own project budget. She assigns a 10-year lifetime to any alternative chosen to meet the goal. The rational for this is simply that most capital equipment will fall into the IRS's 10-year lifetime category. Next, she determines cost reductions resulting from reducing solvent usage by 80% (approximately $100,000). She realizes these are overestimates because there may be some increases due to higher costs of new products, new wastes, and increased utilities usage or processing time. The overestimate provides an absolute maximum price limit for the purchases of new equipment and is desirable in helping to make a first cut through the options. Now the engineer subtracts a portion of this value to account for taxes. Multiply the $100,000 by 60% (one minus the corporate tax rate). If this step was not performed, the budget amount would always be larger than the present value and would not provide the desired ROI. Multiply this value by the present value factor for a 10-year project at the 15% hurdle rate. This would be $60,000 × 5.0188 = $301,128. The engineer knows that the premise of ROI calculations is setting the net present value equal to zero. She also knows that when the net present value is zero, the present value of the incremental operating costs (cost savings) equals the capital costs (purchase of new equipment) after taxes. Therefore, her operating budget for new equipment should never exceed $301,000 if the company's 15% ROI must be met.

What are the constraints in this equation? First, the not-to-exceed budget figure is derived from the savings in operational expenses. The greater the savings, the larger the budget could be and still yield the minimum 15% ROI. Second, there may be ancillary cost savings outside the production process that have not been quantified by this first estimate. Additional cost savings would allow a larger up-front cost with the same rate of return. If these cannot be known during budget development, they will result in an increased rate of return on the project, improving the likelihood of selection and implementation. Third, in this case, the lifetime of the alternative was arbitrarily set at 10 years. The allowable budget would change if this factor changed. A shortening of the lifetime to 5 years would reduce the not-to-exceed budget amount to $201,132. Increasing the lifetime to 15 years would increase the budget to $350,844. Managers and engineers can choose more comfortably an anticipated lifetime by using ACRS depreciation tables. These tables define four basic lifetimes allowable under current IRS tax law and are summarized as follows:

- *3 years:* For cars, light-duty trucks, and machinery and equipment used in research or experimentation.
- *5 years:* For all personal property, most production equipment, and some public utility property.

- *10 years:* For most public utility property and depreciable real property such as buildings and structural components.
- *15 years:* For some public utility property and 15-year property types.

Using these tables will remove some of the ambiguity around how long a piece of equipment might reasonably last, thereby increasing the accuracy of the TCA.

CHAPTER 5

ACTUALLY DOING IT: THREE LEVELS OF TCA DETAIL

THE NEED FOR THREE LEVELS OF DETAIL

The previous chapters have described all the steps needed to generate data for the final calculation, what formulas are used in the calculation, and how to use compound interest tables to shorten the work. This chapter will show by example three levels of TCA that can be adapted to the individual needs and size of the company. Returning again to the central idea of creating a practical tool for measuring pollution prevention benefits, we find one approach to calculating the TCA is not adaptable enough to the wide range of situations. A small business needs a time-efficient tool with a minimum of complexity, one that may allow consecutive, more detailed analyses if the project demands them. This need arises from the limited resources of a small business, which limits the amount of time that can be spent learning a new approach as well as the time it takes to analyze a single project. A larger corporation may have more resources available, such as full-time staff dedicated to costing or environmental management. A larger corporation is more likely to have more intricate projects, with larger numbers of material inputs, cost categories, and variables. In this case, a more detailed analysis would be required. The TCA may even require the use of a spreadsheet to track all the categories and variables. In the middle are all the rest of the potential users who desire expediency and thoroughness, clarity and accuracy. This chapter will present three modes of TCA calculations designed to fit each of the three situations just outlined. The identical example calculation then will be performed using each approach and the benefits and shortcomings discussed.

THE SHORTCUT TCA

The shortcut approach is a no-frills, almost "back-of-the envelope" calculation that yields financial numbers in the shortest time period and with the least amount of data gathering involved. Essentially, this approach looks at the largest cost categories first, removes the effects of income tax, reducing the number of steps in the net present value calculation to a minimum. This approach can be streamlined further if the manager/engineer wants to make certain assumptions about the company's cost of capital and the project lifetime. In some cases, it is possible to multiply the cost savings from the largest cost category by a single decimal number and obtain an approximate present value for the project. If this number is larger than the capital cost of the project, then the manager/engineer knows the project will at least return the initial investment before its projected life expires. If set up properly, and all assumptions clearly explained, this approach can be performed almost as quickly as a payback analysis.[1]

When Is the Shortcut Approach Appropriate?

This level of detail is appropriate for first-cut analysis when savings appear obvious and only a rough approximation is required to prove it. When a company first initiates a pollution prevention program, many "low-hanging fruit," or obvious cost savings, will present themselves. These "no-brainers" often have cost savings three to ten times their purchase costs.

For example, a machine shop has a small electroplating line in the back room, where the rinsewaters discharged to the sewer consistently contain too much zinc. The plant manager desires to appease the local sewer authority and decides to install a series of dragout tanks rinses to decrease the amount of zinc leaving the process. The shop's foreman estimates the company can save $2,500/year on zinc purchases through reusing the dragout tank solution

[1]The shortcut approach was adapted from a study performed for the Pacific Northwest Pollution Prevention Research Center, Seattle, Washington, 1995.

and decreasing the amount of water purchased from the town. The fabrication of the new tanks will cost only $800. In this case, the investment is recovered in less than a year, compliance with the local sewer authority is achieved without penalties, and chemical purchases are reduced.

A good rule of thumb is to use the shortcut approach when the savings from the first two or three cost categories are larger than the initial capital investment. You must know which categories are the most significant for their particular process. In the previous example, zinc purchases and water usage were the major cost categories for the process. Water use decreased significantly as a result of the project, but the decrease in zinc usage would not have been expected unless the foreman knew that the dragout solution could be reused as plating tank makeup water.

Why bother to run a TCA on a project that so obviously saves money and has a less than 1-year payback? Well, the project may have to compete with other capital purchases for approval. A quick analysis could put this project on the top of the list or narrow the choices between many projects that will require additional review. In summary, this level of detail is best used for simple projects, stand-alone processes, or projects having no other competing capital purchases.

When Is the Shortcut Approach Not Appropriate?

The shortcut level of detail is not appropriate in the following situations:

1. *The project impacts several operational areas of the facility.* A project involving many areas of one facility will also involve inputs of energy, chemicals, and labor from a variety of areas. Too many "significant" cost categories turns the shortcut analysis into a more extended TCA, the exact opposite of why one would use it in the first place. Choosing just one area out of all those affected by the process would bias the analysis and exclude costs that may improve or disprove the project's viability. In either case, that additional information is needed and should not be omitted in the rush to save time. Large processes may also have several waste streams or generate wastes that are particularly expensive to dispose. If the analysis excluded the area that generates this waste, a significant risk and liability would fail to be accounted for. Again, this may boost the likelihood of the project being approved or made a priority.

2. *The process has a large number of material or chemical inputs.* A complex process with large numbers of primary or ancillary materials inputs will require too much time to gather all cost data. Each input's contribution to total materials costs is likely to be unknown in the beginning of a TCA. The time involved to gather cost data for the analysis increases as additional costs are added. If a process has 20 material inputs, the shortcut approach ceases to be a time saver if all 20 inputs have to be determined. Estimating or approximating their contribution introduces errors. This compounds the inaccuracies created during the net present value calculation. The analysis of large complex processes deserves careful application of all observational powers and a methodical approach. Shortcuts cannot offer the level detail required. All of these inputs may be of equal significance, and all must be included, or accuracy of the TCA will be compromised. The same reasoning applies that any of these inputs could make or break an analysis and cannot be excluded.

3. *Compare many similar projects to each other.* It is sometimes necessary to develop a series of scenarios or possible options as variations on one major process change. These options may differ only slightly in the type of technology used, or brands of chemicals purchased, or differences in trade-ins, rebates, or other financing available. In this case, either the capital or operating costs may differ enough that a more thorough analysis is required to find the maximum benefit to the company. Sometimes, the bottom-line totals may be identical, yet individual cost categories could vary a great deal. These variations may have impacts on other areas of the manufacturing process that would be overlooked if a shortcut approach were used. A dramatic increase in electricity costs may be masked by a similarly dramatic decrease in chemical purchase costs. This increase in electricity usage could place a strain on the facility's electrical system, causing problems in other areas when operating.

Who Would Use the Shortcut Approach?

The shortcut approach can be used by anyone desiring a quick answer. As will be shown in the next section, the shortcut approach can be structured to minimize the amount of data gathering and calculation required and still yield a present value sum. Those who might use this method most often would include the following:

- Plant managers preparing annual budgets
- Salespeople or consultants introducing new technologies
- Owners of small businesses with relatively simple manufacturing operations

- Pollution prevention teams trying to eliminate projects that will definitely not earn a profit
- Someone in a meeting needing a quick first-cut number for discussion.

How to Use the Shortcut Approach

Reducing a full TCA into a short and simple analysis requires some assumptions and predetermined parameters. These assumptions must be grounded in factual experience to ensure production of a reasonable final figure. If you are not already familiar with the company's capital budgeting process, now would be the appropriate time to develop a deeper understanding of its requirements. The worst possible scenario is to be in a large meeting of peers and present a proposal based on figures that everyone knows are unacceptable by the CFO.

Hurdle Rate Review Chapter 3 for more information on the terminology used here such as "hurdle rate." A company's hurdle rate is usually known by only a few top-level managers. It may be possible that the hurdle rate is more of a "hurdle range" than one fixed number. If this is the case, choose the high end of the range and keep in mind that this makes it more difficult for the project to earn a profit. A common assumed hurdle rate is 10%.

Project Lifetime All projects must have a designated point in time after which they will either be replaced or discarded. Most of the time, this is also the lifetime if the asset is depreciated for tax purposes. Because there will not be any depreciation or tax calculations used in the shortcut approach, this figure is only needed to determine the number of periods in discounting the annual savings. If multiple projects will be compared using the shortcut approach, it is vital that all projects have the same lifetime. Otherwise, the manager/engineer will be comparing apples to oranges. A common assumed lifetime is 10 years.

Write down your company's hurdle rate and the anticipated project lifetime on a scrap of paper and turn to the back of this book (Appendix A, "Compound Interest Tables"). These tables contain all the factors that a single annual number can be multiplied by to obtain a present value for the specified length of time. Across the top row are all the interest rates. For this calculation, consider these rates to be the *hurdle rate.* Down the left side are the periods from 1 to 100. For this calculation, consider these numbers to be the *lifetime* of the project. Select the appropriate hurdle rate and lifetime and follow the corresponding row and column to where the two intersect. This is the compound interest factor that will generate a present value for a the chosen hurdle rate and desired lifetime. Record this decimal on the same scrap of paper and return to this chapter. For the assumed values of 10% at 10 years, the factor is 6.1446.

Step 1. Gather Capital Cost Data Consult the vendor or sales brochure for the purchase price of the equipment. The price without tax will suffice, and unless large modifications are required to install this device, the installation cost can be neglected. Large modifications, for the purposes of the shortcut approach, would be changes such as installing new water lines and waste-treatment plumbing in an area of the plant that never had a water supply before, raising the roof of the building to accommodate an extra tall piece of equipment, or stabilizing or reinforcing floors or roofs to support equipment that exceeds the building's designed weight load. Add the purchase cost of the equipment and any significant installation costs together for a total capital cost.

Step 2. Gather Operating Cost Data Carefully examine the existing manufacturing process. It may be instructive to actually walk around the process machinery and take notes. What are the major inputs and waste outputs? Are they chemicals, materials, or utilities? Look for some of the key indicators listed in Table 5.1 that signal the significance of an input.

Once the largest significant inputs and annual costs have been determined for the current process, place them in a single column on the left side of a page and total them. This total represents an approximate operating cost for the current system (see Table 5.2).

Next, review any literature, vendor information, or price estimates for the proposed process. Compare the input levels of the current system with the proposed system. This step in the TCA requires more estimating and calculating than the previous input analysis because the equipment has not been purchased or installed. Table 5.3 provides questions that may help determine inputs for a system that has not been purchased yet. Look for similar indicators to determine the significant inputs and wastes for this system.

List the inputs of the new system in a second column, to the right of the first column of costs, and calculate the total annual operating cost for the new system (see Table 5.4).

TABLE 5.1 Key Indicators of Input Significance

Key Indicator	May Signify . . .	Need to Check . . .
Large number of pallets of a particular material waiting to be used	This process requires a large amount of this material for production.	Annual purchase records for this material.
Large number of electrical devices such as motors, blowers, rectifiers, pumps, heat or curing lamps	Electricity usage is a large portion of the operating cost of this process.	Utility bills and estimate the fraction of total usage this process represents.
Many drums of a single type of chemical	This chemical is frequently added to the process and used in large amounts.	Annual purchase records for this material.
Large-diameter pipes leading to the process carrying water or other fluids	Water or fluid flow into the process is considerable, but not necessarily continuously flowing.	Specifications of pumps that deliver fluids, or find another measure of fluid flow.
High voltage or other warning signs	Electricity usage may be a large operating cost for this process, and do not touch anything!	Utility bills and any ratings or meters that may reveal electricity usage.
Numerous steam leaks or vapors rising from solutions	Steam or other type of process heat may be a large operating cost.	Proportion of total plant heating or fuel use related to this process.
Large stacks, ducts, or ventilation apparatus over process equipment	Electricity to operate the ventilation system and chemical being removed by ventilation may be significant.	Check proportionate utility usage and purchase records for the chemical being used that evaporates.

(Continued)

TABLE 5.1 *Continued*

Key Indicator	May Signify . . .	Need to Check . . .
Process equipment appears exceptionally clean	Large amounts of chemicals or labor are expended to maintain the equipment or there are cleanliness specifications for the process.	Check purchase records for cleaning chemicals and ask maintenance staff about labor expended on cleaning this process.

TABLE 5.2 Current Significant Operating Costs

Chemical A	\$20,000
Chemical B	\$10,000
Electricity	\$12,000
Total	\$42,000

Step 3. Calculate the Present Value of the Cash Flow The next step calculates the annual costs/savings of the new system that occurs for each year of the project's life. This *incremental cash flow* will then be converted to a present dollar amount. Subtract the annual operating cost of the new system from the annual operating cost of the old system. From the data in Table 5.4, the following incremental cost is determined:

Current operating cost, \$42,000/year − new operating cost, \$30,000/year
= \$12,000/year

The difference, the *incremental cash flow*, is either a positive or negative number, and results in one of three possible outcomes.

1. *The number is significantly negative; no further calculation needed.* For the shortcut approach to be *significantly negative* means the new operating cost total is *at least* 15% greater than the existing operating cost total. When this is the case, the proposed project will never return a positive cash flow or meet the expected hurdle rate. The project may still have redeeming qualities, such as significant reductions in chemical use or pollution, but it will not be a financially attractive investment.

■ **TABLE 5.3 Questions about New System's Inputs**

Question	Answer	Calculation Required
What is the main source of energy for the system?	If energy is replacing a chemical, then energy may be a significant cost.	Estimate annual utility usage.
Does the new system use different chemicals or materials?	If these are replacements for existing chemicals or materials, they may become new significant costs.	Determine the current market cost of the input material and estimate annual usage.
Is the new system more complex in its operations?	If it is, there may be more maintenance involved or a need to replace parts more often.	Discuss this with the equipment vendor and ask for estimated maintenance costs.
Does the new system require fewer line operators?	If so, there may be a substantial savings in production time.	Vendor information should detail processing capcaity and labor needs.
Does the new system generate a new type of waste?	If so, this waste may be more difficult to dispose.	Check unit price for new waste and estimate quantities to be generated by the system.

2. *The number is slightly negative or slightly positive.* In this case, the operating costs for the new process would be ±15% of the current system. This situation calls for a more in-depth analysis of the costs. In essence, the shortcut approach cannot be relied on here. Adding other cost categories will increase the accuracy of the TCA and provide a more reliable estimate of financial performance. It has been found that the additional, smaller categories can influence the shortcut TCA by approximately ±15%. Because of this, it is recommended that a more in-depth analysis be performed.

3. *The number is significantly positive; proceed to next step.* When the incremental costs are *significantly positive,* that is, the new operating cost is

TABLE 5.4 Comparison of Annual Operating Costs

Item	Current	New
Chemical A	\$20,000	\$0
Chemical B	\$10,000	\$0
Chemical C	\$0	\$25,000
Electricity	\$12,000	\$5,000
Total	\$42,000	\$30,000

less than the current operating cost by at least 15%, then the project may be financially feasible. The positive result means that every year the new system will cost less to operate than the current system. The next question to answer is whether it will pay for itself before its lifetime expires.

For the example in Table 5.4, the incremental cost was a positive \$12,000/year, which is much larger than 15% of the current operating cost (\$42,000/year × 15% = \$6,300/year). Therefore, the present value of this incremental cost can be determined.

Step 4. Calculate the Net Present Value Take the incremental cost determined in step 3 and multiply it by the compound interest factor recorded before step 1. By using the numbers from the Table 5.4 example:

Incremental cost, \$12,000 × Factor, 6.1446 = Present value, \$73,735

What does this number mean? It means that over the 10-year life of the project, the company can expect to save \$73,735 in present dollar value. To determine the net present value (NPV), subtract the expected capital cost of the proposed project from the present value. No capital cost was given for this example, which allows several outcomes to be explored.

1. *The NPV is significantly negative; project is not profitable.* A negative NPV means the investment does not pay for itself by the time its life expires. If a capital cost of \$100,000 were assigned to the Table 5.4 example, the NPV would be −\$26,264. The company would actually lose over \$26,000 in the 10 years the project was running. Even though it saves the company money every year, the project does not save the company enough money per year to offset the loss in value of the dollar over the 10-year period.
2. *The NPV is within ±15% of the capital cost.* If this were the result, the uncertainty of the shortcut method would again require further analysis. Incor-

porating more cost categories will drive the NPV toward one extreme or the other and away from uncertainty. If after these iterations were performed and the result was still within the 15% range, a footnote should be added to the summary. This note should explain that external variables such as fluctuations in market prices for input materials may influence the somewhat marginal profitability of the project. In the Table 5.4 example, the calculated PV would be marginal if the capital cost were greater than $62,700. This would yield a NPV less than 15% of the capital cost.

3. *The NPV is significantly larger than the capital cost.* This scenario results in a profitable project. The larger the NPV, the more rapidly the project pays back the initial investment. The return on investment (ROI) can be calculated by an iterative process or by using a financial calculator. Turn to Chapter 6 for an explanation of ROI, why it is used, and how to calculate it using the iterative process.

Shorter Shortcuts

There may be a need for an even faster calculation than the one described. If a salesperson is explaining the benefits of a new technology to an environmental manager and makes claims about the money saving features of the new machine, a shorter shortcut can be used to mentally check the claims. Suppose a manager has a solid understanding of the current operating costs of the production system. She knows that chemical and waste costs are the significant cost categories for the current process, and average about $40,000/year. The salesperson introduces the new technology and claims a 10% reduction in chemical and waste costs. Furthermore, the introductory price for this little gem is only $30,000. The savvy manager mentally calculates the incremental cost savings to be $4,000/year. She has already memorized the compound interest factor normally used for TCAs (10 years at 10% = 6.1446). A quick multiplication of 6 × $4,000 = $24,000 tells her that this technology probably will never pay for itself. As a confidence check, she takes 15% of the capital cost ($30,000) = $4,500 and adds it to the present value of $24,000. The project is still in the red by $1,500. Her interest in this technology has by now substantially waned.

The key to mastering the shortcut approach is to know what the current operating costs are for the processes in the facility. When this is accomplished, the calculations become as fast as those for a payback analysis and much more accurate. The strength of this approach lies in delivering a quick answer on projects that have large positive or negative incremental costs and net present values. As the incremental cost and net present value tend toward zero, the accuracy of this approach fades and requires a more detailed look into the feasibility of the project. The final section of this chapter will analyze a single project using the shortcut approach and compare it with the other two approaches.

THE BALANCED TCA

Occasionally, a project may require more detail or a closer look but there is no time to gather every bit of data. This second approach may be able to help.

What Is the Balanced Approach?

The balanced approach is a compromise between speed and accuracy. It can function as a second review of a project that appeared to be marginal using the shortcut approach. This balanced approach may also appeal to those companies that normally require a fair level of detail in all financial reviews. The main differences between the shortcut and balanced approaches are the number of cost categories included and the use of tax shield calculations. These two differences create a more accurate picture of the financial performance of a project and follow many common accounting practices.[2]

Whereas the shortcut approach reviewed only the largest cost categories, this approach looks at five separate categories for capital costs and seventeen categories for operating costs, as shown in Table 5.5. These categories are arranged in a table for easy reference and can be printed out when performing a TCA.

The balanced approach utilizes the annuity calculation format, similar to the shortcut, to reduce the number of steps required to calculate the present value, and incorporates depreciation and tax calculations.

When Is the Balanced Approach Appropriate?

Use the balanced approach when time allows and/or a more detailed picture of the project's profitability is required. It combines the benefits of being a manageable calculation that does not require a computer, and the flexibility to allow the user to change parameters and compare results using dif-

[2]The balanced approach and associated tables were developed by Mitchell Kennedy, Pollution Prevention Cooperative, 1993.

■ **TABLE 5.5 Balanced Approach Cost Categories**

Capital Costs	Operating Costs		
Equipment purchase	Chemical purchases	Ancillary chemicals	Storage space
Disposal of old process	Waste-treatment chemicals	Waste testing	Waste disposal
Research & design	Safety training	Safety equipment	Insurance
Initial permits	Fees or taxes	Paperwork labor	Annual permits
Building process changes	Production cost	Maintenance labor	Maintenance materials
	Water usage	Electricity usage	Gas/oil/steam usage

ferent interest rates, lifetimes, or tax rates. This approach handles projects easily spanning several operational departments or having many material inputs. The user can set up the TCA table to view two or three different projects side by side to compare differences in specific cost categories or present values. The table structure produces a well-documented analysis that is both easy to read and reproducible by a reviewer.

Who Would Use the Balanced Approach?

This middle-of-the road approach is flexible enough for many applications.

- Small business owners who desire a better-than-average level of accuracy in financial projections
- Environmental engineers and managers at medium-sized to large companies
- Pollution prevention or quality teams trying to select one or more advantageous projects
- State or university technical assistance providers trying to help businesses with environmental issues
- State environmental enforcement officials evaluating a proposed supplemental environmental project (SEP)

How to Use the Balanced TCA

Begin by assembling cost data just as in the shortcut approach. With 22 cost categories to review, this initial step takes considerably longer. Review the cost gathering techniques outlined in Chapter 2 to minimize wasted effort. It is quite possible that the project being analyzed will not have all the data elements that the current process contains. This lack of data is common when new technologies are evaluated. Look for pilot test data, other applications or facilities that currently use the product being evaluated, and communicate extensively with the vendors. For data that cannot be located, decide either to omit it from the TCA or estimate what the cost might be. In either case, include an explanatory footnote in the closing remarks. Once the data for all the categories have been collected or determined not to exist, record them in tables similar to Tables 5.6 and 5.7. This format clearly shows all cost categories for the TCA and allows a side-by-side comparison of the costs for the current and proposed systems.

The worksheet shown in Table 5.6 details the five capital costs included in the balanced TCA. The following narrative will assist the manager/engineer in filling them in for the first time.

Equipment Purchase The cost to purchase all the needed process equipment should be entered here. This would include all materials to build the process, such as filters, lamps, fans, ductwork, piping between process tanks, and other machinery or parts that were not part of the original process.

■ **TABLE 5.6 Capital Costs Worksheet**

Capital Costs	Current Process	Alternative A	Alternative B	Alternative C
Equipment purchase	N/A			
Disposal of old process	N/A			
Research & design	N/A			
Initial permits	N/A			
Building/process changes	N/A			
Total capital costs	N/A			

■ **TABLE 5.7 Operating Cash Flow Worksheet**

Operating Cash Flow		Current Process	Alternative A	Alternative B	Alternative C
Chemical purchases					
Ancillary chemicals					
Storage space					
Waste management	Chemicals				
	Testing				
	Disposal				
Safety Training/ equipment					
Insurance					
Fees & taxes					
Filing paperwork time					
Annual permits					
Production costs	% Inc./Dec.				
	$/year				
Maintenance	Time				
	Materials				
Utilities	Water				
	Electricity				
	Gas/Steam				
Total Annual Operating Cash Flow					

Disposal of the Old Process Equipment such as old vapor degreasers may require cleaning or decontaminating prior to disposal. These costs plus the cost of removing the equipment from the facility should be included. Occasionally, this cost will be a revenue if, for example, a scrap metal dealer offers to pay the company for the value of the metal machinery.

Research and Design Including this category as a capital cost often raises the ire of accountants because it is not always clear-cut whether research and design is a depreciable expense. Keep in mind that this is a decision-making tool, not a method for federal tax preparation. As such it is important to know how much time, labor, and resources went into the development of a new process. This is especially true when several technologies are to be evaluated against each other as well as against the current process. In that case, an off-the-shelf technology that performs as well as one that took 6 months of research to design will have clear financial benefits. For this reason, the research and design costs are included in the balanced TCA process.

Initial Permits More often than not, an alternative technology requires an additional environmental permit because it adds an additional waste discharge or adds a new type of waste not covered under the current permit. The cost for these permits has to be included as part of the capital cost.

Building and Process Changes A new process may require additional electrical supply, water feed lines, or compressed air. It also may require a connection to the plant wastewater treatment system. Installing these services can require cutting through the concrete floor and burying a line or pipe, moving portions of walls, constructing a new circuit-breaker panel or other costly activity. These costs become more important when multiple projects are evaluated against each other. For example, fitting a new process into a room may require moving a wall or enlarging a doorway. If process machine A could be disassembled and brought into the room in pieces, and process machine B could not, there would be a financial benefit to purchasing process machine A. Table 5.6 provides a line to account for this benefit.

Total Capital Costs After filling in all the data for the capital cost table, the total capital cost can be determined. This figure will be divided by the project's lifetime to determine annual depreciation. This straightline method of depreciating the asset simplifies the calculation. The manager/engineer could incorporate other forms of depreciation, but would have to calculate the present value for each year of the project separately, and then total the present values (see the discussion of depreciation methods in Chapter 4).

Operating Cash Flows

Table 5.7 can be used to organize the operating cash flows for the current and proposed processes. These cost categories were reviewed in Chapter 4 and will not be repeated here.

By using this table to organize the operating costs of current and proposed projects, the manager/engineer can tell quickly if all projects have the same cost categories filled in. Estimating the waste-disposal costs for one project and ignoring them for another project will result in inaccuracies. The user must be consistent and either leave the category blank or estimate waste-disposal costs for both projects. Once all the cost categories have been filled in, the total operating cash flow can be determined. For the current process, this number will be a cash outflow, or a cost to the company and is symbolized by placing the number in parentheses. The proposed projects also will be cash outflows unless the project generates a substantial revenue in one or more of the cost categories. A large payment by a scrap metal dealer or other type of recycling bonus could offset the cash outflows and result in a small cash inflow. It is more common for the operating cash flows for the proposed projects to just be smaller cash outflows than the current process.

Cash Flow Summary Worksheet

The balanced approach uses a third worksheet called the cash flow summary worksheet (see Table 5.8). This table guides the manager/engineer through each step of the tax and net present value calculations. It is useful to have this guide if TCAs are not performed on a regular basis or if all three approaches are used regularly.

■ TABLE 5.8 Cash Flow Summary Worksheet

Cash Flow Summary	Current Process	Alternative A	Alternative B	Alternative C
Total operating costs				
Incremental cash flow				
− Depreciation				
Taxable income (TI)				
Tax = tax rate × TI				
Net income				
+ Depreciation				
After-tax cash flow (ATCF)				
PV = ATCF × factor				
Total capital cost (TCC)				
NPV = PV − TCC				
Equivalent Annual Annuity				

The calculations required for this approach may seem intimidating at first. After several example runs, though, most people grasp the operation as nothing more than an inflated income tax calculation. This section of the balanced approach will be demonstrated by an example.

Balanced Cash Flow Summary Example

Step 1. Capital Costs A screw machine company wants to buy a custom-built aqueous parts cleaning system for $10,500. There will not be any other capital costs. See Table 5.9.

Step 2. Operating Cash Flows The cash flows for both the current system and the proposed aqueous system are calculated.

CURRENT DEGREASER'S OPERATING CASH FLOWS

Chemical Use: The company's use of TCE (the solvent) last year was 23,148 pounds. At current market prices for TCA, the company spends $25,000 to purchase the solvent.

Waste Disposal: Current disposal costs are $1,000/year for seven drums.

Fees & Taxes: Use of TCE requires the company to file SARA 313 form R and a state toxics use report. The state collects an annual toxics use fee of $500.

Paperwork: The associated paperwork takes 20 hours for both forms and another 5 hours for permits associated with chemical use.

Maintenance: The vapor degreaser requires a complete clean out every 6 months, taking one worker 3.5 hours each time. The company's labor rate is $20/hour.

Utilities: Electricity usage could not be obtained as the degreaser was not metered separately and specifications on the unit could not be found.

■ **TABLE 5.9 Capital Costs**

Capital Costs	Current Degreaser	Aqueous System
Equipment purchase	N/A	$10,500
Total	N/A	$10,500

EXPECTED OPERATING CASH FLOWS FROM THE AQUEOUS CLEANER The new aqueous cleaning system has a 10-year lifetime.

Chemical Use: Chemical usage for the new system is estimated to be 0.5 gallon/week. The cost per gallon for the new chemical is slightly higher than the TCE's current market price ($14.00/gallon of new cleaner versus $12.50/gallon for TCE), but less is used and it lasts longer. The chemical is not used at levels that trigger regulatory thresholds for toxics use regulations.

Waste Disposal: There are no costs.

Fees and Taxes: There are no costs.

Paperwork: There are no costs.

Maintenance: The new unit requires monthly cleaning, taking one worker just over 2 hours to complete. Metal filings that accumulate at the bottom of the new cleaning tank will be periodically removed and sold as scrap, but no cost data are available. In-line filters are changed every 6 months and discarded as solid waste.

Utilities: The cleaning tank's rinses will consume roughly $10/year of water. Compared to water used in other shop processes, this is a nominal impact. The company heats both the old and new systems with in-plant steam. Cost differences for other utilities are considered marginal.

Table 5.10 shows how each of the operating costs compare. Using the balanced approach allows for quick comparison between two or more projects.

Step 3. Cash Flow Summary Now that the cost data have been entered into the tables, they can be totaled to find the total capital costs and total annual operating costs. Calculating the present value begins with accounting for the effects of depreciation and the tax savings from the use of depreciation.

FIND THE INCREMENTAL CASH FLOW Subtract the total operating costs of the aqueous cleaner from the annual operating costs of the TCA vapor degreaser. See Table 5.11.

FIGURE THE TAXABLE INCOME Divide the total capital costs by the number of years of expected lifetime for the equipment. In this case, divide $10,500 by

■ TABLE 5.10 Operating Cash Flow Worksheet

Operating Cash Flow		TCA Degreaser	Aqueous System	Incremental Cash Flow
Chemical purchases		($25,000)	($400)	$24,600
Waste management	Chemicals	0	0	0
	Testing	0	0	0
	Disposal	($1,000)	0	$1,000
Safety training/ equipment		($20)	($20)	0
Insurance		N/A	No change	0
Chemical use fees		($500)	0	$500
Filing paperwork time		($400)	0	$400
Annual permits		($100)	0	$100
Production costs	% Inc./Dec.	0	0	0
	$/year	0	0	0
Maintenance	Time	($140)	($480)	($400)
	Materials	0	0	0
Utilities	Water	N/A	($10)	($10)
	Electricity	Same	Same	0
	Gas/steam	Same	Same	0
Total Annual Operating Cash Flow		($27,160)	($910)	$26,250

■ TABLE 5.11 Cash Flow Summary

Cash Flow Summary	TCA Degreaser	Aqueous Cleaner
Total operating cash flow	($27,160)	($910)
Incremental cash flow		$26,250

10 years. The annual depreciation of the new equipment is $1,050/year. Subtract this from the incremental cash Flow. See Table 5.12.

DETERMINE THE AFTER-TAX CASH FLOW This example assumes a corporate tax rate of 40%. More accurate calculations could be made with an actual tax rate. Multiply the taxable income by the tax rate and subtract the income tax. Add the annual depreciation to the net income to determine the after-tax cash flow. See Table 5.13.

■ **TABLE 5.12 Taxable Income**

	TCA Degreaser	Aqueous Cleaner
– Depreciation	N/A	($1,050)
Taxable income		$25,200

■ **TABLE 5.13 After-Tax Cash Flow**

	TCE Degreaser	Aqueous Cleaner
Income tax (40%)	N/A	($10,080)
Net income	N/A	$15,120
+ Depreciation	N/A	$1,050

■ **TABLE 5.14 Net Present Value**

	TCE Degreaser	Aqueous Cleaner
Present value (6.1446)	N/A	($99,358)
Total capital costs	N/A	$10,500
Net present value	N/A	$88,858

FIND THE NET PRESENT VALUE Multiply the after-tax cash flow by a present value factor. This factor is determined by the lifetime of the equipment and the company's discount rate. Compound interest tables in Appendix A provide a factor of 6.1446 for a 10-year lifetime at a 10% discount rate. Subtract the total capital costs from the present value to obtain the net present value of the project. See Table 5.14.

Step 4. Financial Viability The annual operating cash flow for running the old degreaser is $27,160. The total annual savings, after taxes, from the new system are $16,169. A simple comparison of capital costs to after-tax savings shows the new unit will pay for itself in less than 1 year. However, the aqueous cleaner is expected to last 10 years, which yields a net present value of $88,858. Table 5.15 shows the previous calculations grouped together for clarity.

The ratio of capital invested to that returned can be found by dividing the present value by the total capital costs. For this example, the ratio is 9.46:1. This means the investment in an alternative to TCE will return $9.46 for every $1 invested. The company will save $1,400/year on avoided paperwork

■ **TABLE 5.15 Cash Flow Summary**

Cash Flow Summary	TCE Degreaser	Aqueous Cleaner
Total operating costs	($27,160)	($910)
Incremental cash flow	N/A	$26,250
− Depreciation		($1,050)
Taxable income		$25,200
Income tax (40%)		($10,080)
Net income		$15,120
+ Depreciation		$ 1,050
After-tax cash flow		$16,170
Present value (6.1446)		$99,358
Total capital cost		($10,500)
Net present value		$88,858

and fees, yet maintain environmental compliance through eliminating regulated substances. If the manager/engineer desired to compute the ROI of this project, a similar calculation could be performed as was done in the short-cut example using iterations. The compound interest factor to search for would be $16,169/$10,500 = 1.5399. The highest compound interest table figure available was for 50% = 1.9965, so the ROI is greater than 50%.

THE ADVANCED TCA

When rigorous study of one option is desired or many alternatives are being considered, the advanced approach may be most useful of the three TCA approaches described.

What Is the Advanced Approach?

The third approach to TCA analysis provides even more detail than the other two. Through extensive research and calculation of present values using a computer spreadsheet, a person can manipulate and calculate more values and variables than with a calculator, pencil, and paper. The advanced TCA can be designed to display several depreciation methods (straight-line, double declining balance, and double declining balance switching to straight line), and include automatic ROI, IRR, and payback calculations (see Chapter 6 for an explanation of ROI). The computer also allows the user to choose a particular year and see how the investment looks during that time period.

It is relatively simple to customize a Lotus 1–2–3 or Microsoft Excel spreadsheet into tables that allow you to enter the data.[3] It is important to note that the primary differences between the advanced and the balanced approaches lie in the computation of the net present value and not in data gathering. The manager/engineer will still have to spend time gathering data, perhaps even more, when using the advanced approach. Larger cost categories such as purchased equipment are broken up into subcategories (delivery, sales tax, spare parts) and these or any other category can be expanded by simply adding additional rows. Extra spaces between rows and blank spaces for additional items can be deleted so that it will fit on one page. Worksheets would be needed for operating costs and for a cash flow calculation sheet showing the present value calculations over the chosen time period. Although there are quite a few data elements to enter, any that are not relevant can be left blank. These worksheets can also be linked together so that following sheets can take data from previous ones, minimizing duplicative data entries.

When Is the Advanced Approach Appropriate?

The additional detail and calculations of the advanced approach fits well in an organization that has high expectations of its cost engineers. The advanced approach is quite useful for highly detailed analysis or for examining incremental adjustments to one particular project. Using the computer, you can adjust the cost of capital by a full or half percentage point, add in an expected inflation effect, or change the purchase or operating costs of multiple subcategories and have the computer recalculate the NPV. The advanced approach also can contain additional information that upper-level management may consider important, such as the ROI, modified internal rate of return (MIRR), and calculations on loan repayments. The beauty of working in a spreadsheet software package such as Excel or Lotus 1–2–3 is that all the formulas and compound interest factors are preloaded in the software and the

[3]The advanced approach was inspired by U.S. EPA P2 Finance software originally developed by the Tellus Institute, 1993.

calculations are handled by the computer. You are encouraged to develop your own spreadsheet application. Once you have an understanding of the software and of TCA, creating a highly customized worksheet will take less than an hour. The example in this approach uses Microsoft Excel Version 5.0. The formulas required for annuity-based calculations are shown in Table 5.19 near to the calculations. Some of the other functions performed by Excel are listed in what follows and can be added to the spreadsheet with a single line.

Depreciation Functions	***Cell Formulas***
Straight line	SLN (*cost, salvage, life*)
Declining balance	DB (*cost, salvage, life, period, month*)
Double declining balance	DDB (*cost, salvage, life, period, factor*)
Variable declining balance	VDB (*cost, salvage, life, start, end, factor, no_switch*)
Sum-of-the-years-digits	SYD (*cost, salvage, life, period*)

Investment Functions	***Cell Formula***
Present value	PV (*rate, number of periods, payment, future value, type*)
Net present value	NPV (*rate, inflow1,inflow2, . . . , inflow29*)
Future value	FV (*rate, number of payments, present value, type*)
Internal rate of return	IRR (*values, guess*)
Modified internal rate of return	MIRR (*values, cost of funds,reinvestment*)[4]

The advanced approach works best when comparing only one alternative to the current process. Two main drawbacks to the advanced approach make comparing multiple options cumbersome.

1. Each alternative requires consideration of all cost categories measured in the current process. Not all cost categories need be filled, however; each one at least must be viewed and considered initially. The example operating cost

[4]Excel functions from the Help Menu and from The Cobb Group, *Running Microsoft Excel 5 for Windows*, Redmond, WA: Microsoft Press, 1994, pp. 498–510.

■ TABLE 5.16 Comparative Number of Data Elements in TCA Approaches

Cost Categories	Shortcut Approach	Balanced Approach	Advanced Approach
Capital costs	1 to 2	5	36
Operating costs	5 or less	16	36

worksheet contains 36 subcategories for current process costs, each of which must be considered initially. Each alternative also must be evaluated on these 36 cost categories for consistency. Again, not all the categories may be needed or filled in, however; there is time involved in reviewing the categories and judging their applicability. There are an additional 36 ***capital cost*** categories to be considered for each alternative. In contrast, the balanced approach has 16 operating cost categories and five capital cost categories, and the shortcut approach has no more than five operating cost categories and one purchase cost. Obviously, the more categories there are, the more time it will take to finish the analysis.

2. Comparison of alternatives against each other is cumbersome. Having many categories and subcategories generates many rows of figures. Standard computer monitors can show only 20 rows of text using a 12-point font size before the user must scroll up or down to view the remaining lines. The number of columns is likewise limited. In the default opening of Microsoft Excel, the user can only view seven columns of data before the screen must be repositioned. These limitations are manageable when working on the computer, as scrolling is simply a matter of pushing buttons or moving the mouse. When printing however, these additional rows are often cut off or placed on separate pages, forcing the user to shuffle papers or tape all the pages together.

Analyzing multiple options on a single worksheet becomes challenging. It requires careful layout of all data elements and ensuring that the right cells are linked to each other across tables or worksheets. Again, the limits of the printed page intrude when a paper copy is desired. A five-option analysis may take five sheets of paper across and two or three down. Reducing the font size just makes it more difficult to read.

Year-by-Year Data Viewing

Spreadsheets also can be created to show each cost category and its value over time, as well as the year-by-year calculations of the after-tax cash flows and net present value. This creates a very impressive display of numbers, but is of questionable merit given the trade-off of having so many sheets of paper to shuffle through. *Always keep in mind that TCA is a means to reach a decision, not an end of and by itself.* It is the final number that is desired.

Who Would Use the Advanced Approach?

As mentioned earlier, this approach could be used in situations where many iterations and subtle differences need to be explored. A project engineer or designer may use TCA in this way. A chief financial officer may adjust costs and capital allocations to meet budget goals. Her support staff of analysts may need TCA to calculate and recalculate profit projections from a pollution prevention project. Anyone with a degree of computer savvy and knowledge of spreadsheet software could customize easily a spreadsheet for his or her own facility or department.

How to Use the Advanced Approach

The first step in performing an advanced TCA is to create a software spreadsheet(s) for the analysis. The one shown in Table 5.17, was constructed in less than 1 hour. There are many cost analysis software packages commercially available. If you choose to use one of these, make sure the cost categories can be edited and added to by the user.

Invest the time up front to thoroughly identify potential cost categories. Set a goal for a total number of categories and subcategories for the process being analyzed. After the initial analysis is complete, other cost categories may be added or deleted depending on the characteristics of the next process being analyzed. A list of potential categories is provided in Table 5.18 as a reference and to underscore the differences in the three approaches. Note that each succeeding approach includes all the cost categories from the previous ones.

After developing the initial spreadsheet, collect the data as in the other two methods and enter them into the computer. If all the functions are properly keyed, the spreadsheet should do all the calculating. The following example uses the Excel spreadsheets in Tables 5.24 and 5.25. If desired, the spreadsheet can furnish other decision criteria such as the internal rate of return or the modified internal rate of return. A payback calculation also could be included.

■ **TABLE 5.17 Spreadsheet Computation of NPV**

Cost Worksheet	Cost Figures	Cell Formula
Lifetime (years)	10	value from previous sheet
Old ann. oper. CF	$75,000	value from previous sheet
New ann. oper. CF	$40,000	value from previous sheet
Incremental oper. CF	$35,000	cell = B5 – B6
Less depreciation	($10,000)	cell = (B3/B4)
Taxable income	$25,000	cell = B7+B8)
Less income tax	($10,000)	cell = – (B9*0.39)
Net income	$15,000	cell = B9+B10
Add depreciation	$10,000	cell = B3/B4
ATCF	$25,000	cell = B11+B12
Present value	$153,614	cell = – PV(10%,10, B13)
Capital cost	($100,000)	value from previous sheet
Net present value	$53,614.18	cell = B14+B3
Calculate IRR		
Capital cost	($100,000)	value from previous sheet
Present value	$153,614	cell = –PV(10%,10,B13)
IRR	54%	cell = IRR (B19:B20)

Extension of the Advanced Approach

Average knowledge of spreadsheet software allows easily automating the TCA, allowing the program to run all three approaches as needed. In this case, the cell formulas would first calculate the net present value using only the costs specified in the shortcut approach. If the answer fell within the uncertainty range, it could prompt the user to enter additional data and then proceed to calculate a balanced TCA or advanced TCA. If the spreadsheets were imbedded in Visual Basic or another OLE programming language, the whole TCA could be made very user friendly. This automated TCA routine then could be linked with a larger decision analysis or project planning program such as Expert Choice or the P2EDGE. In very large corporations, the corporate environmental department could provide a TCA intranet webpage for use by all divisions. This has the added advantage of standardizing the analysis protocol and providing a single point of contact for collecting TCA information on environmental projects.

TABLE 5.18 Cost Categories for the Three Approaches to TCA

Approach	Shortcut	Balanced	Advanced
Capital Costs		**All of the previous plus . . .**	**All of the previous plus . . .**
	Equipment purchase	Disposal of old system	Working capital
		Initial permits	Delivery and freight charge
		Building modifications	Training
		Process modifications	Start-up chemicals
		Research & design	Salvage value
Operating Costs		**All of the previous plus . . .**	**All of the previous plus . . .**
	Chemicals	Ancillary chemicals	Waste storage
	Materials	Storage	Recycling revenues
	Waste Disposal	Waste testing	Semiannual planning
		Waste-treatment chemicals	5-year permit renewals
		Safety training	Self-audits
		Personal protection equipment	Outside certifications
		Insurance	Quality control reject costs
		Chemical taxes and fees	Impacts on other processes
		Annual permits	
		Paperwork time	
		Production costs	
		Maintenance labor	
		Maintenance materials	
		Water/sewer use	
		Electricity	
		Gas/oil/steam	

A COMPARATIVE EXAMPLE

The following example shows the strengths and weakness of each of the three approaches discussed earlier. This example was specifically tailored to illustrate certain aspects of each approach.

Example Background

The Norte Canning Company desires to replace its label printing press with a high-speed inkjet printer. This new technology from Germany will be attached to the canning line and prints the label as the cans come off the line and onto the conveyor to the packaging machines. Its link with the production computer allows instantaneous changes to the type of label and label design as the change in the type of can or can contents hit the conveyor. It will eliminate all solvent-based inks at the plant. The environmental manager in charge of evaluating the process talked with the vendor and obtained the specifications on feed rates, chemical usage, maintenance time, energy consumption, and other factors. Table 5.19 shows the shortcut TCA.

The shortcut revealed a marginal project, that is, one whose present value is within 15% of the capital cost of the new equipment. The engineer decides there must be other benefits to the project that are not being captured and now performs a balanced TCA. Table 5.20 shows the capital costs associated with the project.

Note that the old process is scheduled to be sold for scrap and there is some welding and fabricating required to install the new process over the existing canning line. In all, the balanced approach added 3.7% in cash flows to the capital costs.

The operating costs as shown in Table 5.21 have dramatically changed. Several key benefits to the system were not considered in the shortcut

■ **TABLE 5.19 Shortcut Approach**

Costs	Old	New	Difference
Purchase cost	NA	$40,000	($40,000)
Ink purchases	$30,000	$28,000	$2,000
Waste disposal	$ 8,000	$ 2,500	$5,500
Total operating cost	$38,000	$30,500	
PV factor = 4.8			
(7 years @ 10%)	4.8 × $7,500 = $36,000		

TABLE 5.20 Capital Costs for Label Machine

Capital Costs	Current Process	New Label Machine
Equipment purchase	NA	$40,000
Disposal of old process (sold as scrap)	NA	($500)
Research & design	NA	0
Initial permits	NA	0
Building/process changes	NA	$1000
Total capital costs	NA	$40,500

TABLE 5.21 Operating Costs Using the Balanced Approach

		Current Process	New Label Printer
Chemical purchases		($30,000)	($28,000)
Waste management	Chemicals	0	0
	Testing	($300)	($100)
	Disposal	($8,000)	($2,500)
Safety training/ equipment		($300)	($100)
Insurance		NA	No change
Chemical use fees		0	0
Filing paperwork time		($5,300)	($1,200)
Annual permits		($400)	($200)
Production costs	% increase/ decrease	0	Uncertain
	$/year	0	Uncertain
Maintenance	Time	($3,000)	($1,000)
	Materials	($500)	($400)
Utilities	Water	NA	($300)
	Electricity	($1,300)	($2,000)
	Gas/steam	N/A	N/A
Total annual operating cash flow		($49,190)	($35,800)

■ **TABLE 5.22 Cash Flow Summary Using Balanced Approach**

Cash Flow Summary	Current Process	New Label Printer
Total operating costs	($49,190)	($35,800)
Incremental cash flow	NA	$13,390
− Depreciation	—	($5,714)
Taxable income (TI)	—	$7,675
Tax = tax rate * TI	—	($3,070)
Net income	—	$4,605
+ Depreciation	—	$5,714
After tax cash flow (ATCF)	—	$10,319
PV = ATCF * Factor		
(7 years 10% = 4.8684)	—	$50,239
Total capital cost (TCC)	—	$40,000
NPV = PV − TCC	—	$10,239
Equivalent Annual Annuity	—	NA

analysis because they seemed less significant than chemical purchases and waste disposal. It turns out that the solvent carrier in the existing ink forces the company to comply with several time-consuming environmental regulations, run exhaust vents over the printing process, and considerable maintenance due to the buildup of ink residues in the trays and rollers.

Table 5.22 shows how the balanced approach draws out additional cost savings and highlights the projects true pollution prevention potential.

Now the project has a net present value of positive $10,000. This is larger than the 15% margin of error, so the engineer could stop the analysis at this point, except that he knows his corporate managers will want better figures. With this in mind, he completes the advanced TCA by refining some the data already gathered and plugging in applicable cost categories from Table 5.21. The engineer realizes this new technology will require some training prior to start-up and chooses to include all the freight and delivery charges for the new unit. He also includes the impact of this technology on other processes. This impact consists of eliminating a backlog of canned products whenever the old label printer had to change over to a different label. The backlog decreased total output of canned product by 3% and total revenues by $10,000. This backlog was added as a cost to the old system, and a value of zero was added to the new system because the new technology eliminates the backlog.

The spreadsheets used for this analysis were designed on Microsoft Excel 5.0 and are slightly more complex than the one shown in Table 5.16, but use the same cell formulas and financial analysis functions included with the software. Developing all four worksheets for this example required less than 8 hours of effort. The first spreadsheet allows the user to enter the critical values for such facility-specific variables as cost of capital, inflation rate, project depreciation period, income tax rate, and wage rates for maintenance and environmental compliance staff. Keeping these figures up front separates them from other data elements, so they can be found quickly if they have to be changed. This sheet also contains summary information about the profitability of the project. Having the final analysis displayed beneath the assumptions that went into the analysis allows for a quick review and adjustment if warranted.

A review of Table 5.23 shows the project has a 10% cost of capital, a 0% inflation rate, and a 7-year lifetime. The company's tax rate is 40%, and its fully loaded hourly rate for the maintenance staff is $13/hour and $21/hour for the environmental staff. Table 5.24 starts the TCA calculations for the total capital cost of the new label printer. In Table 5.24, you can see where several categories have been expanded (compared to the balanced approach worksheets) to include additional costs such as delivery charges, state sales tax, engineering work, and training for the line workers on how to operate the new printing system. Also note the greater level of detail in itemizing the costs

■ TABLE 5.23 TCA Spreadsheet Package

Please enter the following facility-level variables:			
Discount rate	10.00%	Maintenance staff hourly wage—	
Inflation rate	0	fully loaded ($/hr)	13
Project lifetime	7	Environmental staff hourly wage—	
Corporate tax rate	40%	fully loaded ($/hr)	21
Proceed to Worksheets 2 and 3 and return here when done			

Financial feasibility of selected project:					
Net present value		**ROI**		**Payback =**	1.79
At year 3 =	$ 2,075	At year 3 =	13%	(In Years)	
At year 6 =	$29,362	At year 6 =	33%		
At year 10 =	$37,430	At year 10=	35%		

■ TABLE 5.24 Capital Cost Spreadsheet for Label Printer Analysis

Capital Costs	Cost	Totals
Equipment purchases		
New label printer	$37,736	
Delivery charge	$500	
CT sales tax (6%)	$2,264	
Stocked spare parts	$200	$40,700
Engineering & design		
In-house engineering	$100	
		$100
Training		
Pilot runs to establish tolerances	$500	
Line operator training from vendor	$1,000	
		$1,500
Utility upgrades of new systems		
Electricity (220 V)	$0	
Freshwater feed	$100	
		$100
Disposal of old process/equipment		
Sold old printing presses for scrap	($500)	
		($500)
Building/process modifications		
Additional writing	$200	
Reinforcement of canning conveyors	$400	
		$600
Salvage value		
	$0	$0
Total capital cost		$42,500

within each category. These details help identify costs that are larger than average and services that might be better done in-house or by a contractor.

Through his persistent efforts, the engineer refined the anticipated capital cost of the project, and it rose from $40,500 to $42,500.

In reviewing the operating costs in Table 5.25, it can be seen that every cost category will have a cost reduction under the new printing process, except for utility usage. The largest reduction was initially overlooked and only discovered after considering the impacts of the new technology on other

TABLE 5.25 Operating Costs Spreadsheet for Label Printing Process

Current Printing Process			New Label Printing Process			
		Total			Total	Incremental Cost
Chemicals/materials inputs			**Chemicals/materials inputs**			
Inks	$30,000		Inks	$28,000		
		$30,000			**$28,000**	**$2,000**
Waste Management			**Waste Management**			
Tests	$250		Testing	$180		
On-site handling	$100		On-site handling	$150		
Hauling	$300		Hauling	$150		
Insurance	$0		Insurance	$0		
Disposal	$8,000		Disposal	$2,600		
		$8,650			**$2,900**	**$5,750**
Environmental compliance labor			**Environmental compliance labor**			
Manifests (hours/year)	48	$1,008	Manifests (hours/year)	5	$105	
Reports (hours/year)	190	$3,990	Reports (hours/year)	57	$1,197	
Training (hours/year)	15	$315	Training (hours/year)	5	$105	
Permits (hours/year)	19	$399	Permits (hours/year)	10	$210	
		$5,712			**$1,617**	**$4,095**
Environmental fees & taxes			**Environmental fees & taxes**			
Permit fees	$200		Permit fees	$100		
		$200			**$100**	**$100**

(Continued)

TABLE 5.25 *(Continued)*

Current Printing Process			New Label Printing Process			
Insurance			**Insurance**			
	$0			$0		
		$0			**$0**	**$0**
Production increase/decrease			**Production increase/decrease**			
Changeover backlog	$10,000		Eliminate backlogs	$0		
		$10,000			**$0**	**$10,000**
Utilities			**Utilities**			
Electricity	$1,500		Electricity	$2,000		
Water			Water	$300		
Sewerage			Sewerage	$100		
		$1,500			**$2,400**	**($900)**
Maintenance Costs			**Maintenance Costs**			
Labor (hours/year)	300	**$3,900**	Labor (hours/year)	100	$1,300	
Materials	$500	**$4,400**	Materials	$400	**$1,700**	**$2,700**
Total		**$70,074**	Total		**$39,634**	**$23,745**

production processes. The plant expects to reduce its backlog of cans that develops every time the current printing press stops to change printing plates and inks for a new product run. This backlog creates shortages farther down the production line, with the end effect being a reduction in total output capacity from the plant. (If how this progression of events limits plant production capacity is unclear, refer to Chapter 10 for the section titled the "Theory of Constraints.") Resolving this capacity bottleneck results in a virtual increase in plant capacity without the addition of more canning equipment. Overlooking such production-related opportunities is common when performing the shortcut approach; in the rush to obtain an answer, the more subtle but nonetheless important opportunities are missed.

The final worksheet contains a year-by-year analysis of all cost categories and present value calculations. This allows the engineer and reviewers to quickly identify the period when the project begins to turn a profit. This area could be expanded to include interest payments on financed capital (i.e., bank loans). Most process-level improvements are currently made using the company's equity because lending institutions have become wary of loaning money to companies for environmental projects. This is no small barrier to the implementation of pollution prevention projects that can be overcome only through education of bankers and lenders to show them that pollution prevention reduces liabilities and preserves the environment.

Although the first spreadsheet gives summary information on the financial feasibility of the project, the fourth spreadsheet (Table 5.26) allows the user to see each calculation and the effects of the cost of capital and inflation in each year of the project. The project has a highly favorable net present value ($37,430) and return on investment (35%), and a payback of 1.79 years. Also of interest in Table 5.26 is the difference in allowable depreciation effects between the straight-line and the double declining balance switching to straight-line methods. As described in Chapter 2, the double declining balance method allows the user to allocate more of the total value of an object in the early stages of its life than at the end of its life. The cell formula used to decide when to switch from declining balance to straight-line depreciation of the remaining balance is

```
=IF(I2>'1'!C9,0,IF((('2'!D28-SUM(D17:H17))/('1'!C9-
H2))>D.B.('2'!D28,'2'!D27,'1'!C9,6),(('2'!D28–SUM(D17:H17))/('1'!C9–
H2)),D.B.('2'!D28,'2'!D27,'1'!C9,6)))
```

This formula is included to save the programmer time if and when it is decided to construct a TCA spreadsheet. The cell and sheet numbers would have to be changed to fit each application's format.

The values for the capital costs, operating costs and NPVs for each of the three approaches are listed in Table 5.27. Examining this table reveals the relative accuracy of the three approaches.

Overall, it is important to notice the trend of increasing costs and an almost paradoxical increase in savings (greater NPVs). The payback period for the project also is shortened as would be expected when savings increase and the capital cost remains relatively stable. As the calculation pulled in more data, the numbers showed the large differences in the two printing processes. In the first approach (shortcut), the analysis results in a negative NPV, but one that is within the 15% range, requiring further calculations. Stopping at this point might lead a manager to ignore the potential savings identified using the other approaches. However, the calculation required less than 2 minutes to perform once the initial data had been gathered.

By using the balanced approach, the true benefits of this project come forth. Intuitively, the concept of eliminating a solvent from a printing operation should save money as well as the environment. The balanced approach reaffirms this by exposing the large cost savings resulting from reduced environmental compliance costs. This calculation required significantly more time to gather data (approximately 6 hours) and another 20 minutes to fill in the tables by hand and run the calculations on a pocket calculator. The greater the level of cost detail currently collected by a facility, the shorter the time required to gather data.

The third approach delivers the greatest detail, both in terms of overall reliability of the numbers and the display of calculations and assumptions. The large increase in incremental cash flow is directly related to the production capacity increase from eliminating bottlenecks. This savings is compounded over the life of the project and results in a present value savings almost twice the capital cost. The ROI for the project is automatically calculated by Excel to be 35%, probably three times the company's hurdle rate. Gathering cost data for this approach required 4 hours beyond the balanced approach, for a total of 10 hours. The actual calculation was practically instantaneous, but would require 20 minutes to enter all the data. However, once the data have been entered, adjustments to the cost of capital, project lifetime, or other variables will result in a new set of values without further time expenditure.

TABLE 5.26 Cash Flow Summary Spreadsheet for Label Printing

Year (Period)	0	1	2	3	4	5	6	7	8	9	10
Inflation effect	1.000	1.000	1.000	1.000	1.000	1.000	1.000	1.000	1.000	1.000	1.000
Operating Costs											
Chemical/material inputs	2,000	2,000	2,000	2,000	2,000	2,000	2,000	2,000	2,000	2,000	2,000
Waste management	5,750	5,750	5,750	5,750	5,750	5,750	5,750	5,750	5,750	5,750	5,750
Environmental compliance labor	4,095	4,095	4,095	4,095	4,095	4,095	4,095	4,095	4,095	4,095	4,095
Environmental fees & taxes	100	100	100	100	100	100	100	100	100	100	100
Insurance	0	0	0	0	0	0	0	0	0	0	0
Production increase/decrease	10,000	10,000	10,000	10,000	10,000	10,000	10,000	10,000	10,000	10,000	10,000
Utilities	(900)	(900)	(900)	(900)	(900)	(900)	(900)	(900)	(900)	(900)	(900)
Maintenance	2,700	2,700	2,700	2,700	2,700	2,700	2,700	2,700	2,700	2,700	2,700
Total operating costs	23,745	23,745	23,745	23,745	23,745	23,745	23,745	23,745	0	0	0
Depreciation Effects											
Tax depreciation (by straight-line)		6,071	6,071	6,071	6,071	6,071	6,071	6,071	0	0	0
Tax depreciation (by DDB switching to SL)		12,143	8,673	6,195	4,425	3,688	3,688	3,688	0	0	0

(Continued)

■ TABLE 5.26 (*Continued*)

Cash Flow Summary												
Total operating costs			23,745	23,745	23,745	23,745	23,745	23,745	23,745	0	0	0
− Depreciation (DDB/SL)			12,143	8,673	6,195	4,425	3,688	3,688	3,688	0	0	0
Taxable income			11,602	15,072	17,550	19,320	20,057	20,057	20,057	0	0	0
− Income tax at:	40.0%		4,641	6,029	7,020	7,728	8,023	8,023	8,023	0	0	0
Net income			6,961	9,043	10,530	11,592	12,034	12,034	12,034	0	0	0
+ Depreciation (DDB/SL)			12,143	8,673	6,195	4,425	3,688	3,688	3,688	0	0	0
+ Salvage value			0	0	0	0	0	0	0	0	0	0
After-tax cash flow		(42,500)	19,104	17,716	16,725	16,017	12,034	15,722	15,722	0	0	0
Present value @ 10%		(42,500)	17,367	14,642	12,566	10,940	7,472	8,875	8,068	0	0	0
Net present value at period = n			($25,133)	($10,491)	$2,075	$13,015	$20,487	$29,362	$37,430	$37,430	$37,430	$37,430
		$1	−55%	−9%	13%	24%	29%	33%	35%	35%	35%	35%

■ TABLE 5.27 Comparison of Values for Each of the Three TCA Methods

Cost Category	Shortcut TCA	Balanced TCA	Advanced TCA
Capital costs	$40,000	$40,500	$42,500
Incremental cash flow	$ 7,500	$13,390	$23,745
Net present value	($ 4,000)	$10,239	$37,430

CLOSING

Each of the three approaches has clearly defined strengths and weaknesses. The environmental manager must decide when each approach is appropriate given the time constraints and level of detail required for each project. As with anything with repeated use, the data gathering and calculation steps become familiar and faster. Using a materials tracking program and a TCA spreadsheet, an experienced environmental engineer easily can produce an advanced TCA in less than 2 hours.

CHAPTER 6

OTHER MEASURES OF PROFITABILITY

CHAPTER CONCEPTS

This chapter explores the various ways to measure the profitability of one investment (project) against another. The end of the chapter focuses on measures of project profitability: the internal rate of return (IRR) and the equivalent annual annuity (EAA) that are typically used in conjunction with the net present value (NPV). These additional measures may be used in an analysis along with the net present value (NPV) to show the results in a form more familiar than NPV or to allow the comparison of two very different projects (the apples-to-oranges issue). The ability to recognize when each one of these two measures is needed will add credibility to your TCA and prevent misusing them.

WHY IS MORE THAN ONE MEASURE NEEDED?

You will need these other measures to increase the accuracy of the comparisons between complex projects or very dissimilar projects. TCA uses the present value and net present value to arrive at a dollar value of the project. Because more detailed analyses like balanced or advanced TCA typically involve more complex systems or multiple-project analysis, it is likely that additional comparisons will be needed to clarify the results. The more detailed TCAs are also used typically when the project is competing with other capital purchases or investment options. In these cases, it is necessary to compare all projects at a glance. The following paragraphs summarize the most common measures, their benefits, and limitations.

Do companies use more than one decision measure? You bet. The measure used and the amount of time that more than one measure is used can be gleaned from the following three surveys. Kaplan and Atkinson[1] surveyed 367 Fortune 500 companies in 1985 revealing that 78% used either internal rate of return or net present value as a primary tool. In 1988, Pike and Wolf[2] demonstrated that 54% of British companies used more than one decision measure (such as IRR and NPV). Finally, in 1994 EPA[3] commissioned a survey on environmental cost accounting practices in which over half of the 149 respondents stated they use some form of discounted cash flow analysis in the intial screening stages, final evaluation stage, or both.

SIMPLE TOOLS

Payback Analysis

This tool is commonly used for simple projects or as a quick check on profitability. Simply put, payback is the amount of time, usually years, required to pay back the initial investment. The shorter the payback, the more attractive the project. This conclusion flows from the assumption that a faster return of the invested capital allows for reinvestment and minimizes the loss of value due to inflation. The payback period is calculated quickly by dividing the total capital cost of the project by the project's annual savings. The majority of people who do payback analyses expect them to be under 3 years.

Payback's Benefits and Limitations The two main advantages of the payback analysis are that it requires only two numbers and the quick speed at

[1]R. S. Kaplan, and A. A. Atkinson, *Advanced Management Accounting*, 2nd ed., Englewood Cliffs, NJ.: Prentice Hall, 1989.

[2]R. N. Pike, and M. B. Wolfe, *Capital Budgeting in the 1990s. London:* Chartered Institute of Management Accountants, 1988.

[3]U.S. Environmental Protection Agency, *Environmental Cost Accounting for Capital Budgeting: A Benchmark Survey of Management Accounting*, #EPA742-R-95-005. Washington, DC: U.S. EPA Office of Pollution Prevention and Toxics, 1995.

which it can be calculated is short. There are, however, serious shortcomings to using this tool, which were discussed in detail in Chapter 5. These weaknesses have to do with the tool's inability to capture revenues occurring beyond the point of payback. Payback is suitable for small projects having large savings. When conducting complex TCAs, more sophisticated and more accurate tools are needed.

Return on Investment (ROI)

A very similar and frequently used tool is the return on investment (ROI). The yield, or amount to be received above the investment cost, can be expressed as a percentage of the total amount of capital invested. The interest rate that provides this yield is the ROI. The ROI divides the annual net income (gross income minus depreciation) by the capital cost, resulting in a percentage of investment returned for that year, whereas payback divides the capital cost by the incremental cash flow to yield the number of years required to pay back the investment.

Benefits/Cost Ratio or the Profitability Index

The benefits/cost ratio is another simple tool for "back-of-the-envelope" calculations. This tool is also referred to as the profitability index (PI). The present value of all the financial benefits occurring over the life of the project is divided by the total capital cost for the project. Note the difference between payback and the benefits/cost ratio— in the former, the result is a number of years, and in the latter, it is a simple ratio. Note also that the benefits/cost ratio uses the present value of all revenues over the project's lifetime. Favorable projects are those with ratios greater than 1:1. That is, the benefits are at least equal to or greater than the costs of the project.

Benefits/Cost Ratio's Benefits and Limitations This tool offers the advantage of quick calculations through a ratio requiring only simple division of the present value. Because the benefits/cost ratio uses the present value calculation, it is more accurate than the payback analysis.

The main disadvantage of this ratio is the lack of information presented to the user. If you were to look only at the ratio and not the calculations that produced the ratio, there would be no way to know the actual costs and benefits. A project requiring $100,000 of investment capital can have the same benefits/cost ratio as a project requiring only $1,000. Yet, there is a significantly larger impact on the capital reserves of the company if the first project is chosen. This impact restricts the company's ability to invest in other projects.

MORE COMPLEX TOOLS

Net present value, internal rate of return, and equivalent annual annuity are widely accepted as profitability indicators and part of the common language of capital budgeting and managerial accounting. As alluded to in this chapter's introduction, NPV forms the core of the TCA analysis and IRR and EAA would be used when detailed TCAs are performed or there is a need to express the results in a commonly understood measure.

Net Present Value (NPV)

This book uses net present value as the primary tool for comparing alternative investments. Net present value is defined as the sum of all the project's future revenues during the project's life minus the capital cost of the project. The only difference between net present value and present value is the subtraction of the total capital costs in the former. This method is mathematically derived and its application is thoroughly described in Chapter 5.

Net Present Value's Benefits and Limitations Net present values' largest advantage is the conversion of future cash flows into present worth amounts. This allows you to compare two projects over time, adjust for inflation, loan payments, and price fluctuations, which result in a more accurate estimation of the project's worth. The formula for finding the present value also allows you to adjust for the level of risk involved with a project. For example, genetic engineering projects may or may not result in a useable product yet require large capital investments. Therefore, the investment in genetic engineering projects can be seen as highly risky, because the investor may or may not earn a large enough profit (or any profit) on the investment. By comparison, purchasing a new drill press or lathe would be far less risky because the output of the machinery is well known and fairly dependable. The profit may not be as great as with genetic engineering, but there is less likelihood of losing money. When evaluating these two opportunities you can assign different discount rates to them to compensate for the different amounts of risk involved. The genetic engineering project may have a 20% discount rate and the drill press 10%.

The main disadvantage of net present value is the amount of calculating required to produce a number. Each cash flow within the project's lifetime must be converted to present dollar amounts, through either a compound interest formula or compound interest table. These values are then totaled and the total capital cost subtracted. Fortunately, there are many computer spreadsheet programs and even some calculators that can calculate these figures. This decreases the workload but still requires significantly more time than a payback analysis.

Internal Rate of Return (IRR)

Every project represents an investment of either the company's own capital reserves or an amount borrowed from a financial institution. The company hopes this investment will yield a profit. When internally financing a project, companies compare expected profits against profits from a known and secure investment. Corporations often have a minimum acceptable rate of return (MARR), which is the lowest interest rate the company will accept on any investment proposal. MARR is usually the rate paid by an investment opportunity that is always available to the company, such as bonds or mutual funds.

When externally financing the project, the company will expect to earn a profit greater than the interest payments on the loan. The interest rate on the loan is the cost of capital. MARR is always higher than the cost of capital.

The internal rate of return (IRR) is the rate at which the project pays back the initial investment. Or put in another way, it is the rate at which interest is earned on the unrecovered balance of the investment so that no balance remains at the end of the project's life.

The IRR of any investment must be greater than the MARR and the cost of capital in order for the project to be accepted by people reviewing the financial analysis.

How to Calculate IRR Calculating the IRR for a *single payment* involves working the compound interest formula backwards to solve for the interest rate instead of solving for the present value. When working with a *stream of payments*, as is most often the case in a TCA, the solution becomes more complex. This version of the present value formula used for streams of cash flows contains a sigma (summation function) and must be solved by trial and error or by using a calculator programmed to solve for IRR:

$$P = \sum \left[F \times \frac{1}{(1 + i)^n} \right]$$

The trial-and-error method is best explained by example. By using the shortcut approach described in Chapter 5, a previously proposed project (the example in Chapter 5, Table 5.4) was calculated to have a NPV of $73,765 over a 10-year lifetime at a 10% annual interest rate. The CEO of the company does not understand NPV and would like to know the IRR on this project given a $50,000 capital cost.

As shown in Chapter 4, the NPV = present value − capital cost. The IRR would be the point at which all the money invested had been returned, or NPV = 0. The next step is to substitute "0" for the NPV in the NPV equation. Then:

$$\text{Present value} - \text{capital cost} = 0$$

or

$$\text{Present value} = \text{capital cost}$$

The last equation shows that the IRR is also the point at which the present value of the incremental cash flows is equal to the capital cost of the project. This present value is equal to

$$\text{Incremental cost} \times \text{compound interest factor}$$

The IRR is, therefore, the interest rate at which

$$\text{Incremental cost} \times \text{compound interest factor} = \text{capital cost}$$

Here is the iterative method for determining the IRR for the example in Table 5.4 using a $50,000 capital cost (this is an assumed capital cost because the example did not state the actual capital cost).

At the 10-year period where the present value would be equal to the capital cost of $50,000, the compound interest tables can be used to find a factor and corresponding interest rate. The incremental cost is given as $12,000. Substituting these numbers into the equation yields

$$\text{Incremental cost } (\$12{,}000) \times \text{factor} = \text{capital cost } (\$50{,}000)$$

Solving for the factor yields

$$\text{Factor} = \$50{,}000/\$12{,}000 = 4.1667$$

The IRR will be the interest rate at a period of 10 years (which was the given project lifetime) where the compound interest factor equals 4.1667.

From the compound interest tables in Appendix A, the factors along the 10-year period line read as follows:

Percent	***Factor***	***Present Value of $12,000***
10	6.1446	$73,735
13	5.4262	$65,114
16	4.8332	$57,998
19	4.3389	$52,066

The tables in Appendix A go up to only 19%, so the conclusion can be drawn that the IRR is greater than 19%. Is this a valid conclusion?

It is known that an inverse relationship exists between interest rates and present values of annuities. This holds because it is known that the basic present value calculation discounts the value of the annuity by the inverse of (1 + the interest rate). Refer to the compound interest formula given earlier if you require more clarity on this relationship. This relationship is true, so the conclusion can be tested by multiplying the incremental cost by the 19% compound interest factor (4.3389). The value created should be slightly larger than the desired $50,000 because the interest rate (19%) is slightly smaller than desired (remember the inverse relationship of interest rate and present value). The preceding present value table shows this to be true. The actual IRR is around 20.5%.

Internal Rate of Return's Benefits and Limitations Internal rate of return figures are simple to compare because they are always percentage points. These percentages can be compared directly to the company's hurdle rate, the cost of borrowing money, or the rates of return of more secure investments such as bonds. Percentages are also simple numbers and most managers know what an IRR is. For these reasons, the IRR is one of the most commonly used tools for comparing the financial feasibility of projects.

However, IRR is not a panacea for comparison. Its primary limitation is the same as the benefits/cost ratio or profitability index; it does not consider the magnitude of the investments. A $1,000 investment returning $5,000 may have a greater IRR than a $100,000 project returning $200,000. The IRR does not distinguish that the profits from the larger project ($100,000) are 20 times larger than that of the smaller project ($4,000). Exclusive use of IRR for project comparison could result in lost opportunities for greater profit.

IRR also neglects the hurdle rates and discount factors used in NPV analysis. In the example of investing in a drill press (discount factor of 10%) or genetic engineering (discount factor of 20%), the interest rates of each project were adjusted to reflect the relative risk of investing. If each option gave *the same incremental cash flow*, $10,000 for example, over a lifetime of 10 years, the net present values would not be the same. For the drill press, this present value would be $61,446, and for the genetic engineering project, it would be $41,925.

It is possible, however, that the projects could have *the same the capital costs*, $30,000, for example. In this case the IRR on each project would be

$$\text{Drill press IRR} = \$10{,}000 \times \text{factor} = \$30{,}000$$

$$\text{Genetic engineering IRR} = \$10{,}000 \times \text{factor} = \$30{,}000$$

The two equations are identical and yield identical IRRs of just under 30%. Yet there is $20,000 more profit to be made from the drill press investment (as shown by present value calculation), even though genetic engineering is state of the art and more in vogue. The IRR completely misses the difference in the present values.

This discrepancy occurs because the interest rate used in that analysis was intentionally set low due to the low risk of the investment. The cash flows over its life, therefore, are not discounted as highly as the genetic engineering project's cash flows, and result in a larger NPV for the drill press. Remember that the larger the discount rate, the faster the loss of value of future dollars.

Equivalent Annual Annuity (EAA)

There is usually more than one way to improve a process or eliminate a chemical, and all of these options may not have equal lifetimes. When evaluating two or more projects each having different lifetimes, the net present value calculations cannot be compared directly. The following example illustrates this issue.

A heat-treating company treats steel parts by heating them in an oil-fired oven. Smoke and odors from the incomplete combustion of fuel oil and residues on the parts have resulted in complaints by the neighbors and concerns about worker health and safety. The company will either

- retrofit the old ovens to burn natural gas and retire them in 5 years or
- install a new resistance heating technology expected to last 15 years

One of these projects may have a significantly larger NPV than the other, making it appear more profitable. However, the first option will last only 5 years before another investment must be made. The second option probably requires more capital, but will last longer than the first.

Accounting theory states that the project with the shorter lifetime allows the user to reinvest the money more quickly and potentially earn a greater profit. The project's equipment may also require replacement at the end of its useful life. These events would impact the ability of the company to make a profit, and the net present values have to be modified to account for the difference. One method to evaluate two options having different lifetimes is the equivalent annual annuity (EAA).

The EAA approach consists of finding each option's net present value and calculating an amount (an annuity) that provides equal annual payments over the project's life. This is done by dividing the net present value by its present value factor (from Appendix A, "Compound Interest Tables"). The resulting EAAs can be compared to one another or divided by the discount rate to produce the infinite horizon net present value. The infinite horizon NPV is the value of the project under the assumption that the company will replace the equipment continually with an identical one for an infinite number of times.

Consider two projects, one with a 4-year lifetime and a second with a 9-year lifetime; their respective NPVs might look like those shown in Table 6.1. At first glance, the 9-year project appears the most profitable given that it has the larger NPV. Because these two projects have different lifetimes, an EAA is calculated by dividing the NPVs by their respective present value factors (4 years at 10% = 3.1699; 9 years at 10% = 5.7590). The EAAs how the 4-year project to be the most profitable. Dividing the EAA by the discount rate (10%) yields the infinite horizon net present values.

Equivalent Annual Annuity's Benefits and Limitations In order to analyze options of different lifetimes using the EAA method, several assumptions must be made:

1. The projects' equipment will be replaced with identical equipment at the end of each lifetime.
2. Each replacement provides the same cash flows as its predecessor, that is, no change in capital or operating costs.

■ TABLE 6.1 Equivalent Annual Annuity

	4 Years	9 Years
Net present values	$ 3,700	$4,200
Equivalent annual annuity	$ 1,168	$ 729
Infinite net present value	$11,683	$7,292

The flaws in these assumptions are fairly obvious. First, people rarely continue to buy the same product over and over again. Manufacturers often discontinue product lines either because of planned obsolescence or due to lack of demand. Technological changes often make older products obsolete. Examples of technological obsolescence abound, such as the buggy whip and the 8-track tape player. Economic fluctuations might force up the prices of the products beyond what someone is willing to pay for them, or materials costs may force a switch to cheaper less-durable materials, which shorten the lifespan of the new product.

Second, it is equally as likely that economic factors have changed since the purchase of the original equipment and that the cash flows will not be the same. Inflation in the national economy causes a constant loss of value in all products. The rate of inflation changes almost every fiscal quarter. As a general rule, the farther out into the future the analysis extends, the less predictable and less accurate will be the analysis. Following this rule to its logical conclusion means that an infinite horizon NPV should be reviewed carefully and not used as a hard-and-fast number for budgeting beyond the first lifetime.

As an environmental manager, you will need to choose among these comparison tools and draw on their strengths to highlight the important aspects of the project under analysis. Keep in mind that each tool has its inherent flaws built on assumptions of static conditions, short project timelines, or zero technological innovation. The most prudent approach uses several comparison tools in tandem to compensate for these assumptions. You may use both NPV and EAA during a balanced TCA or include payback, ROI, and NPV in the spreadsheet package you develop for an advanced TCA.

CHAPTER 7

CASE STUDIES OF TCA

This chapter shows how TCA has been applied in real situations. The companies covered range from small electroplating job shops all the way to multinational corporations. Each case study describes unique obstacles or aspects to help you learn from the experience of others. All case studies focus on pollution prevention technologies. The names of the companies have been omitted to protect their identities.

Two of the eighteen case studies were presented in Chapter 2, as these projects focused more on gathering and understanding process information and cost data. Therefore, this chapter starts with Case Study 3.

CASE STUDY 3: Machine Maintenance

Step 1: Review of Production Process

A large manufacturing facility used 1,1,1-trichloroethane as a general maintenance solvent to degrease machine parts. The solvent cleans lubricants off parts in need of repair, removes dried-on spilled coatings and other soils from all coating, filling and capping, drying, and testing machines (see Figure 7.1). Each machine operator has a small bucket of solvent that is refilled from a larger drum. This is not an uncommon

■ FIGURE 7.1 Process Flow Diagram for Machine Cleaning

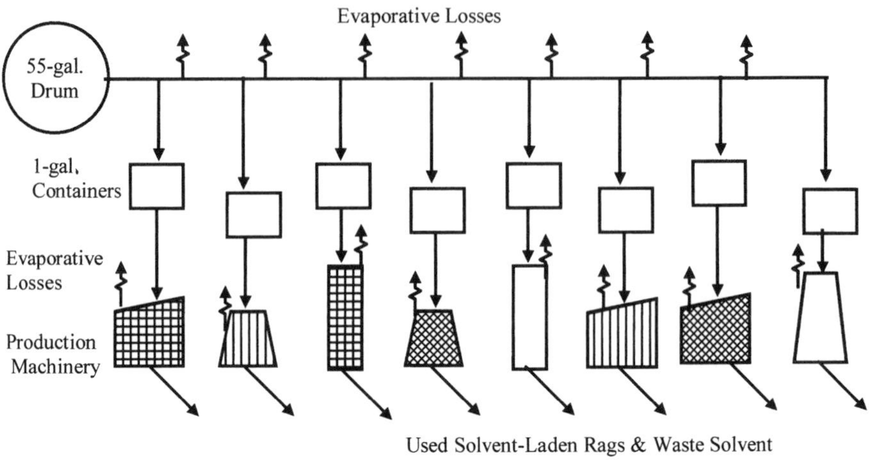

practice in large and small shops all over the country, yet it can cost more than the convenience is worth.

Step 2: Gathering Cost Data

Sources for the data in Table 7.1 include the purchasing department and conversations with the machine operators, maintenance staff, and process engineers. The site walk-through allows us to observe cleaning practices and how the small cans of solvent are stored (covered, not covered, near a fan, etc.).

Because 1,1,1-trichloroethane is used as a maintenance chemical, several cost categories do not apply. The chemical is used entirely for maintenance of the machines and therefore the process of cleaning the machines does not have a maintenance cost attached to it. Also costs connected with production do not apply because maintenance is done when machines are not running. There are no utility costs connected with the use of this chemical as it is applied manually and not used in a vapor degreaser.

1. The costs of purchasing 1,1,1-trichloroethane are significantly below current market rates. On a per pound basis, 1,1,1-trichloroethane is currently selling for $1.39/lb. Purchasing records show a price of $0.61/lb, or 60% lower than market rates. The analysis of any alterna-

TABLE 7.1 Operating Cost for Using 1,1,1-Trichloroethane

Cost Item	Description	Cost Factor	Unit Price	Total Units	Total Cost
Purchase costs	Chemicals			1,741 gal	
	1,1,1-trichloroethane	100%	$ 0.61/lb	(1)	$11,958.00
Storage/floor space	Ancillary storage	0	0	0	0
	Chemical storage	100%	$ 15.50/ft^2	24 ft^2	$ 372.00
	By-product (scrap) storage	0	0	0	0
	Emissions (waste) storage	—	—	—	N/A
Waste management	Treatment chemicals	N/A	N/A	N/A	N/A
	Testing	—	—	—	N/A
	Disposal	—	—	660 gal	$ 1,290.00
Regulatory compliance	Safety/training equipment	—	—	—	Minimal
	Fees & taxes				
	Base fee	14%	$4,625	—	$ 647.50
	Chemical	100%	$1,000	—	$ 1,100.00
	Manifests, Test Reports	—	—	—	N/A
Insurance		N/A	N/A	N/A	N/A
Production costs	$/Unit of product	—	—	—	—
	Labor/unit product	—	—	—	—
	Other	—	—	—	—
Maintenance	Time	N/A	N/A	N/A	N/A
	Materials	N/A	N/A	N/A	N/A
Utilities	Water	N/A	N/A	N/A	N/A
	Electricity	N/A	N/A	N/A	N/A
	Gas/steam	N/A	N/A	N/A	N/A
Total annual operating cost					$15,367.00

tives should account for a potential cost increase in solvent prices as CFCs are phased out and demand for alternatives increases.

2. Disposal costs are also lower than current market rates. Waste disposal manifests and invoices show a disposal cost of $109/drum, whereas national rates are in the $300–400/drum range. Should these disposal prices increase in the short term, the financial feasibility of alternatives will improve considerably. Although the waste solvent is combusted as a fuel, there are still potential liabilities during the transport of the solvent to the company and back out to the treatment, storage and disposal facility.

Long-term liability connected with the use of 1,1,1-trichloroethane is fairly high and is a prominent motivating force to eliminate its use. It is a suspected carcinogen, a groundwater contaminant, and a potential ozone-depleting substance. There was no direct insurance impact due to the use of this chemical.

Step 3: Defining Alternatives

The company's plant manager, process engineer and a consultant developed a short list of alternatives and selected two for further analysis.

Alternative 1: Find a New Cleaning Chemical There are currently hundreds of alternative solvents that are chlorine- and petrochemical-free. Numerous machine shops and maintenance depots have replaced 1,1,1-trichloroethane with a less toxic cleaner. The issues to be dealt with include the type of soils, how they get on the parts, processing time to clean the part, and additional health risks to workers.

Alternative 2: Limit Use Through Worker Training There may be significant reductions possible through working with the maintenance staff and training so as to reduce the use of solvents. Additionally, training workers to keep containers covered and storage drums closed will result in fewer evaporative losses. This type of training can be keyed to an incentive program that gives back a portion of the money saved by better housekeeping practices.

Step 4: Conducting a Balanced TCA

This section and tables detail the financial feasibility of the selected options. "Cost Assumptions" explain the calculations and special conditions found in Table 7.2.

■ **TABLE 7.2 Financial Analysis of the P2 Project**

Capital Costs	Description	Old Process	New Process
Equipment purchase			$9,000
Disposal of old process			$ (48)
Research & design			$ —
Initial Permits			$ —
Building/process changes			$ 432
Total capital costs			$9,384
Operating Costs	**Description**	**Old Process**	**New Process**
Purchase costs	Chemicals	$ 27,531	$ 444
Storage/floor Space	Chemical storage	$ —	$ —
	By-product storage	$ —	$ —
	Emissions storage	$ —	$ —
Waste management	Treatment chemicals	$ —	$ —
	Testing	$ —	$ —
	Disposal	$ 1,320	$ —
Regulatory compliance	Safety/training equipment	$ 20	$ 20
	Fees or taxes	$ 1,100	$ —
	Manifests, test reports	$ 240	$ —
Insurance		N/A	N/A
Production Costs	$/Unit of product	$ —	$ —
	Labor/unit product	$ —	$ —
Maintenance	Time	$ 112	$ 576
	Materials	$ —	$ —
Utilities	Water	N/A	$ 7
	Electricity	Same	Same
	Gas/steam	Same	Same
Annual operating costs		30,323	1,047

(Continued)

■ **TABLE 7.2** (*Continued*)

Cash Flow Summary	Description	Old Process	New Process
Total operating costs		$(30,323)	$ (1,047)
Incremental cash flow			$ 29,276
−Depreciation			$ (938)
Taxable income			$ 28,338
Income tax (40%)			$(11,335)
Net income			$ 17,003
+Depreciation			$ 938
After-tax cash flow			$ 17,941
Present value	10 yr @ 10% = 6.1446		$110,240
Total capital cost			$ (9,384)
Net present value			$100,856

Cost Assumptions

1. The current purchase price for this chemical is 60% below market rates. Other manufacturers would be paying $24,712/yr for the same quantity.
2. The current disposal costs are significantly lower than market rates. Other manufacturers would be paying $3,600/yr for the same quantity.
3. Total annual operating costs are $14,018. Adjusting the disposal and purchase costs for market rates raises this cost to $30,431.
4. Research costs for finding and testing a substitute solvent are based on current staff work of 3 mo × 40 hr/wk × 2 people at $20/hr.
5. This purchase cost is based on a substitution of a mixture combining 50% terpene with 50% mineral spirits. Terpenes are $14/gal and mineral spirits are $2.50/gal. Given a similar annual usage quantity, this would be 773 gal of terpenes ($10,822) and 773 gal of mineral spirits ($1,932). This results in a total annual purchase cost of $12,754.
6. Disposal costs for the terpene–mineral spirits blend are typically lower than chlorinated solvents because of fewer emissions treatments required for fuel blending a chlorinated versus a nonchlorinated solvent. This difference can be as much as 50%.

Financial Analysis Conclusions

This project results in a positive net present value given the data provided. However, market rates for 1,1,1-trichloroethane and disposal of the same will be increasing dramatically as demand increases, supply decreases, and more federal and state chemical taxes are applied. If the project were evaluated using current market rates, it becomes more attractive. The net present value in this case results in an annual savings of $20,000.

CASE STUDY 4: Lost Product Reclamation

Step 1: Review of the Production Process

A beverage producer adds fumaric acid to adjust and standardize the acidity and flavor of the beverage. The concentrations of this chemical in the finished product are very low. The ingredient is added by hand directly to the product in the blending vessel. The fumaric acid becomes soluble and is then processed through the appropriate lines, filters, pasteurizers, and so forth, to the filling containers. See Figure 7.2. Cleaning between production runs results in lost product and lost fumaric acid. As it is a regulated substance, the environmental team desires to increase the efficiency of its use and decrease its loss.

Step 2: Gathering Cost Data

The unique aspects of this case include the small amounts of material added to each batch, lack of other processes or inputs, and lack of utilities. Even though the overall cost contribution of this chemical is small, it is indispensable in the production of the product. Table 7.3 shows costs directly and indirectly attributable to the use of fumaric acid.

1. All costs for this chemical will be orders of magnitude smaller than the purchase cost. This is a direct result of the chemical being only a small part of the total product mix. Fumaric acid makes up less than 0.001% of the total volume of beverages processed. Attributing costs to such a small component results in fractions of dollars and insignificant numbers when compared to the average annual purchase cost of $130,000.
2. Fumaric acids' portion of waste management costs are relatively small. The only time it is discharged to the facility's wastewater treatment plant is during a cleaning session. The calculation, therefore,

■ FIGURE 7.2 Process Flow Diagram for Fumaric Acid Use

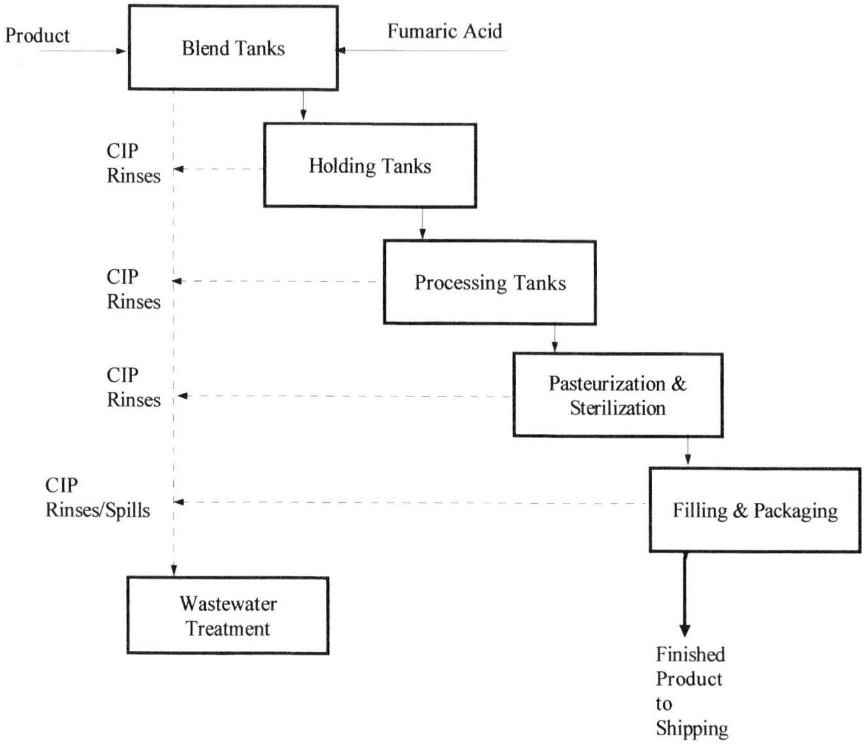

involves first taking a portion of the total waste management costs that reflects cleaning of the processes where fumaric acid is used. Then, a percentage of that number is taken to reflect the amount of fumaric acid in relation to total liquids leaving those processes. The majority of the water treatment is for color, solids, and organics content and is not related to the small quantity of acid that goes into the system.

3. Production costs related to labor involved with the use of fumaric acid were not available. Similarly, maintenance costs could not be proportioned to this chemical. The overall loss rate of 0.4% attests to the current efficient delivery and mixing of this chemical.
4. The company has more than adequate storage space and would not charge back any costs to the fumaric acid processes at this time.
5. Long-term liability connected with the use of fumaric acid is practically zero. Fumaric acid is a derivative of citric acid and is found in

TABLE 7.3 Operating Cost for Using Fumaric Acid

Cost Item	Description	Cost Factor	Unit Price	Total Units	Total Cost
	Chemicals				
Purchase costs	Fumaric acid	100%	$ 1.06	97,329	$103,169
Storage/floor space	Ancillary storage	0%	$ —	0	$ —
	Chemical storage	0%	$ —	0	$ —
	By-product (scrap) storage	0%	$ —	0	$ —
	Emissions (waste) storage	0%	$ —	0	$ —
Waste management	Treatment chemicals	Minimal	$ —	0	$ 50
	Testing	Minimal	$ —	0	$ 50
	Disposal	0%	$ —	0	$ —
Regulatory compliance	Safety/training equipment	25%	$11,180	1	$ 2,795
	Fees & taxes				
	Base fee	25%	$ 4,625	1	$ 1,156
	Chemical	100%	$ 1,100	1	$ 1,100
	Manifests, test reports	25%	$ 460	1	$ 115
Insurance					N/A
Production costs	$/unit of product	0%	$ —	0	$ —
	Labor/unit product	0%	$ —	0	$ —
	Other	0%	$ —	0	$ —
Maintenance	Time	0%	$ —	0	$ —
	Materials	0%	$ —	0	$ —
Utilities	Water	0%	$ —	0	$ —
	Electricity	0%	$ —	0	$ —
	Gas/steam				N/A
Total annual operating cost					$108,435

the biological Krebs's cycle of living cycles. It quickly breaks down into other forms of acids and sugars in the environment, posing no threat of groundwater contamination, cancer, global warming, or ozone depletion. It is used in powdered form; when inhalation is a concern, workers wear masks. Once mixed with water, the acid can pose a reactive risk; however, no other reactives are used during the mixing process. Because of this limited risk potential, insurance costs are not a factor in this analysis and the impact of public perception is also a negligible factor.

Step 3: Defining Alternatives

Alternative 1: Air Purge and Reclaim Liquids This option involves refitting pipelines and straightening bends to allow the use of compressed air to be forced through the pipelines and push out leftover beverage. Expelled product could be collected and reintroduced into the bottling process. This cleaning technology also would decrease the need for flushing pipes and result in less water use. This cleaning method has been used in many food and chemical facilities as a first pass to clear the pipes and recover product. Whereas this project is estimated to reduce fumaric usage by only 0.5%, it will reduce fumaric acid by-products by 39%. A cost analysis of this project follows.

Alternative 2: Dissolve Fumaric Off-Line This option involves creating a premix vessel for fumaric acid and dissolving quantities of the powder in a solution prior to introducing it to the product batch. Current practice allows for undissolved fumaric acid to settle to the bottom of the mixing tank and accumulate. This is later washed from the tank during cleaning or may effect the pH readings of the product as the volume in the tank decreases. Estimates of reductions for this options range between 5 to 10% for total usage. No data exist on exactly how much fumaric acid does not dissolve in the batches and what effect this has on production and cleaning operations.

Step 4: Conducting a Balanced TCA

The environmental health and safety team was eager to see if process changes would result in a profitable venture. Conservative numbers were applied to the TCA to give a worst-case scenario. Table 7.4 shows the conditions applied to the evaluation of the project and the full TCA.

■ TABLE 7.4 Financial Analysis of the P2 Project

Capital Costs	Description	Current Process	Air Purge
Equipment purchase			$ 3,000
Disposal of old process			$ —
Research & design			$ 1,000
Initial permits			$ —
Building/process changes			$ 1,000
Total capital costs			$ 5,000
Operating Costs	**Description**	**Current Process**	**Air Purge**
Purchase costs	Chemicals	$ 103,168	$102,462
	Drums	$ —	$ —
Storage/floor space	Chemical storage	$ —	$ —
	By-product storage	$ —	$ —
	Emissions storage	$ —	$ —
Waste management	Treatment chemicals	$ 50	$ 50
	Testing	$ 50	$ 50
	Disposal	$ —	$ —
Regulatory compliance	Safety/training equipment	$ 2,795	$ 2.795
	Fees or taxes Base chemical fee	$ 1,155	$ 1,155
	Chemical fee	$ 1,100	$ 1,100
	Manifests, test reports	$ 115	$ 115
Insurance		N/A	N/A
Production Costs	$/Unit of product	N/A	N/A
	Labor/unit product	N/A	N/A
Maintenance	Time	$ —	$ —
	Materials	$ —	$ —

(*Continued*)

■ **TABLE 7.4** *(Continued)*

Operating Costs	Description	Current Process	Air Purge
Utilities	Water	$ —	$ —
	Electricity	$ —	$ —
	Gas/steam	N/A	N/A
Annual operating costs		108,433	107,727
Cash Flow Summary	**Description**	**Current Process**	**Air Purge**
Total operating costs		$ (108,433)	$ (107,727)
Incremental cash flow			$ 706
−Depreciation			$ (500)
Taxable income			$ 206
Income tax (40%)			$ (82)
Net income			$ 124
+Depreciation			$ 500
After-tax cash flow			$ 624
Present value	10 yr @ 10% = 6.1446		$ 3,832
Total capital cost			$ (5,000)
Net present value			$ (1,168)

Financial Analysis Conclusions

Despite the reclamation of the product, the annual usage of fumaric acid would still be far above the reporting threshold, requiring the payment of all applicable chemical fees an taxes. The project does not look favorable based solely on reducing the usage of fumaric acid. The net present value in fact is negative, meaning the project will not pay for itself over its lifetime. However, if the savings from reclaiming formerly lost product is added in, the project looks more favorable. Approximately 75,000 gal of product could be reclaimed annually. Even at an assumed cost of $0.50/gal, the project saves over $35,000/yr. Therefore, the project is more important in saving product than in reducing fumaric acid losses.

Unique Aspects of This Case

It is interesting to note the unseen benefit revealed by the TCA. Here the original goal was to reduce the use of a chemical. Although the reductions would not be profitable by themselves, the ability to reclaim product that previously had to be treated as a waste creates cost reductions not seen at the outset. Implementation of this project also may increase the level of control over the process by increasing the accuracy of the pH reading in the mixing tanks. This would be an additional benefit.

CASE STUDY 5: Metal Stamping

This large-sized company manufactures metal casters from plastic, stainless steel, and cold- or hot-rolled steels. These materials are stamped, tumbled, and assembled into a finished product. In trying to meet multiple environment mandates, the company is interested in knowing more about the components of the steel, how the materials flow through the system, and the profitability of any potential improvements. Figure 7.3 shows a generalized process flow diagram for materials inputs and losses.

Step 1: Review of the Production Process

For the purposes of this case study, we will focus on just one of the components of the steel—manganese. Because it is a common component

■ **FIGURE 7.3 Process Flow Diagram of Manganese Usage**

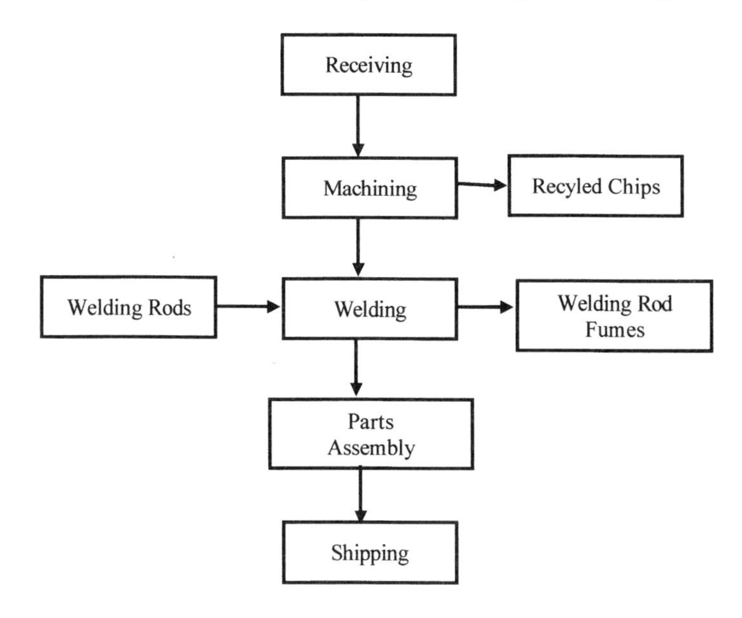

of the steel stock and welding rod, tracking materials flow will cover the entire facility. The manganese comes into the facility in concentrations of 1.5–2.5% in the steels and welding rods. The steel is usually flat stock, rolls, or bar stock, and is then blanked and stamped to rough form. After stamping, the piece is bent or formed to the desired shape in pneumatic and hydraulic presses. Angles and brackets may be welded onto the piece and the entire assembly, and tumbled to remove dirt and rough edges. Plastic wheels are press fitted onto axles and assembled with bearing races and holding brackets into a finished product. Manganese serves as the base metal for the three steels used in this operation: stainless, hot-rolled, and cold-rolled. Stainless steel contains 2.0% manganese, and cold- and hot-rolled steels contain 1.5%. Other metals in varying concentrations make up the remainder of the metal.

Welding is done using a continuous-feed MIG welder. Parts are stamped, cleaned by tumbling, and then any required brackets or supports are welded on. Parts are then either assembled into final form or sent out for plating and then assembled into final form. The welding operation uses welding rods containing 2.5% manganese to aid in heat transfer and surface preparation of the parts being welded.

Step 2: Gathering Cost Data

Our primary contacts for cost information are the purchasing manager and the president. Process information is readily available from shop foremen. The company has a materials inventory system for all product stock. Invoices or receipts were available for most other cost categories. Production and inventory records provided information on welding rod usage. Table 7.5 shows costs associated with using manganese.

1. Storage costs for raw materials are included for the portions of steel and scrap stored on-site until used or picked up for recycling.
2. There were no direct costs to production for the use of this material other than the purchase costs of the steel.
3. The maintenance associated with the use of steels containing manganese consists of sweeping the floors and emptying the machine trays that catch the chips and turnings. These sweepings, chips, and turnings are placed in a roll-off container for recycling. Estimates of

■ TABLE 7.5 Cost of Manganese

Cost Item	Description	Cost Factor	Unit Price	Total Units	Total Cost
Purchase costs	Chemicals				
	Manganese	2%	$ 0.34	6,898,000 lb	$47,540
	Ancillary				
	Welding rod	2%	$ 1.12	10,782 lb	$ 242
Storage/floor space	Steel storage	100%	$ 8/ft^2	800 ft^2	$ 6,400
	Scrap storage	100%	$ 6/ft^2	300 ft^2	$ 1,800
	Emissions (waste) storage	—	—	—	N/A
Waste management	Treatment chemicals	—	—	—	0
	Testing	—	—	—	0
	Scrap recycling	2%	$ 0.03/lb	2,435,938	($1,226)
Regulatory compliance	Safety/training equipment	0	0	0	0
	Fees & taxes	$4,625	$1,100	—	$ 5,725
	Manifests, test reports	—	—	—	N/A
Insurance		—	—	—	N/A
Production costs	$/Unit of product	—	—	—	N/A
	Labor/unit product	—	—	—	N/A
	Other	—	—	—	0
Maintenance	Time	2%	$ 15/hr	79 hr	$ 237
	Materials	—	—	—	N/A
Utilities	Water	—	—	—	0
	Electricity	—	—	—	N/A
	Gas/steam	—	—	—	N/A
Total annual operating cost					$61,944

time for this task are 5 min each day for each machine operator and 1.5 hr/day for a designated maintenance person.

$$5 \text{ min/day} \times 20 \text{ persons} \times 5 \text{ days} \times 50 \text{ wk} \times \$15/\text{hr} = \$6{,}250;$$
$$1.5 \text{ hr/day} \times 1 \text{ person} \times 5 \text{ days} \times 50 \text{ wk} \times \$15/\text{hr} = \$5{,}625;$$
$$\$6{,}250 + \$5{,}625 = \$11{,}875$$

4. Utilities cost connected with the use of steels containing manganese is exclusively the cost of electricity to run the machine that turns out products. This could not be measured or estimated for allocation to the cost of using manganese.
5. Long-term liability connected with the use of manganese is low. It is not considered a groundwater contaminant, carcinogen, mutagen, or teratogen. It is not an ozone precursor or depleting substance. The largest risk comes from inhalation of fumes from welding operations. Because of this limited risk potential, insurance costs are not a factor in this analysis, and the impact of public perception is also a low-weighted factor.

This analysis does not capture the true costs of using the steels and the revenues lost through generating scrap steel. For example, of the total quantity of stainless steel purchased, 33% is recycled. The average purchase cost of stainless is $1.46/lb. Stainless steel scrap is bought at $0.18/lb. Therefore, every pound of stainless steel sent to the recycler costs the company $1.28. Although the recycler paid over $12,000 for some 70,000 lb of scrap, the company lost $77,658.

Mixed scrap reveals a similar cost. Of the 6,700,000 lb of cold-rolled, hot-rolled, and bar stock steel purchased for $0.31/lb, 35% was turned into scrap and sold for $0.02/lb. This results in a materials cost of $631,672.

Step 3: Defining Alternatives

The company is running at over 70% efficiency in usage of incoming metals, yet is losing $709,330 per year through purchase price and scrap resale price differentials. The only way to decrease this loss is to increase the quantity of metal that becomes saleable product.

Alternative 1: Closing Tolerances to Reduce Scrap Generation The manganese usage data point to the largest potential savings available.

Previous calculations revealed a cost savings opportunity valued at $709,000 through increased steel stock usage efficiencies. The company is doing a good deed by selling the scrap stock as a by-product; however, a large annual savings could be achieved by not generating the scrap in the first place.

Closing gaps between parts punched from a single strip or sheet of steel could reduce materials usage and scrap generation. This would require replacing or modifying dies and tools to be more precise without slowing production or affecting quality. It is not possible to achieve 100% materials efficiency, but there is room for improvement between 70% and 100% efficiency. Closing tolerances is estimated to reduce scrap generation by 10%.

Alternative 2: Preventative Maintenance Worn tools, bushings, hydraulic couplings, and other machine parts all contribute to "slop" in the operation of a machine and directly effects the efficiency of materials use. A well-maintained press with a sharp and clean die block will hold closer tolerances for a longer period of time than a dull and pitted die on a poorly maintained press. Preventative maintenance results in time savings, from not having to rework or discard off-spec pieces, and ultimately cost savings. The company estimates an aggressive preventative maintenance program could reduce scrap generation by 5%. This option will be financially analyzed.

Step 4: Conducting a Balanced TCA

The president's keen interest in the identified cost savings demanded a detailed TCA to look at the feasibility of closing operating tolerances. Table 7.6 shows the profitability of the previously stated options.

Cost Assumptions

1. Press modification costs are based on projects at other metal stamping operations. Approximately $10,000 is required to replace dies, piston rings, bearings, and other parts that directly effect the degree of movement during hammer blows. An additional $10,000 is included for installation of miscellaneous equipment, labor, production interruptions, and trial runs.

■ TABLE 7.6 Financial Analysis for the P2 Project

Capital Costs	Description	Current Process	Close Tolerance	Per Month
Equipment purchase			50,000	N/A
Disposal of old process			N/A	N/A
Research & design			N/A	N/A
Building/process changes			10,000	10,000
Total capital costs			60,000	10,000
Operating Costs	**Description**	**Current Process**	**Close Tolerance**	**Per Month**
Purchase costs	TUR Chemicals Mananese	47,540	42,786	45,163
	Ancillary Welding rods	242	242	242
Storage/floor space	Chemical storage	6,400	6,400	6,400
	By-product storage	1,800	1,800	1,800
	Emissions storage	N/A	N/A	N/A
Waste management	Treatment chemicals	0	0	0
	Testing	0	0	0
	Scrap recycling	(1,226)	(1,103)	(1,164)
Regulatory compliance	Safety/training equipment	0	0	0
	Fees (i.e., TURA) Base TURA fee	4,625	4,625	4,625
	Chemical fee	1,100	1,100	1,100
	Manifests, test reports	N/A	N/A	N/A
Insurance		N/A	N/A	N/A
Production costs	$/Unit of product	N/A	N/A	N/A
	Labor/unit product	N/A	N/A	N/A
	Other	0	0	0

(Continued)

■ **TABLE 7.6** ***(Continued)***

Operating costs	Description	Current Process	Close Tolerance	Per Month
Maintenance	Time	237	214	237
	Materials	0	0	500
Utilities	Water	0	0	0
	Electricity	N/A	N/A	N/A
	Gas/steam	N/A	N/A	N/A
Annual operating costs		61,944	57,167	60,067
Cash Flow Summary	**Description**	**Current Process**	**Close Tolerance**	**Per Month**
Total operating costs		$ (61,944)	$ (57,167)	$ (60,067)
Incremental cash flow			$ 4,777	$ 1,877
−Depreciation			$ (6,000)	$ (1,000)
Taxable income			$ (1,223)	$ 877
Income tax (40%)			$ 489	$ (351)
Net income			$ (1,712)	$ 1,228
+Depreciation			$ 6,000	$ 1,000
After-tax cash flow			$ 4,288	$ 2,228
Present value	10 yr @ 10% = 6.1446		$ 26,347	$ 13,689
Total capital cost			$ (60,000)	$ (10,000)
Net present value			$ (33,653)	$ 3,689

2. A 10% reduction in the generation of scrap allows the same number of pieces to be produced with 10% less material. Usage of welding rods would be unaffected by this project.
3. There is potential to reduce the required amount of storage space; however, it is more likely the same quantity of steel will be kept on-site, but fewer deliveries will be made. Therefore, there will not be a significant impact on materials storage costs.
4. Scrap generation will decrease by 10%, resulting in 10% less revenue from the scrap recycler.

5. Unfortunately, this project will not lower manganese usage below regulatory reporting thresholds, and the company will still have to pay state environmental fees.
6. Because the company will be generating less scrap, it will take less time to clean up the production flow and various machines that generate scrap. Therefore, a 10% reduction in the cost of maintenance connected to this operation is included.
7. The preventative maintenance (PM) program entails an initial outlay for organization, training, and development of a schedule and task list. This is estimated to cost $10,000, and is a 10-year investment.
8. The PM program will yield a 5% reduction in scrap generation and translate to a similar reduction in steel purchases.
9. The same assumptions for storage space under the "close tolerances" project are assumed here (see no. 3). Reductions from this project will not be large enough to drop below the reporting threshold for manganese.
10. Maintenance costs for this project will increase because of the need to replace parts more frequently and to spend more time maintaining equipment.

Financial Analysis Conclusions

The project to close tolerances on the punch presses is not financially sound. Over the 10-year life of the project, it will cost an estimated $34,000. The estimates on machinery retrofits are very conservative. The higher they become, the less financially attractive this project becomes. This project would become more viable if reductions in steel usage were higher. These reductions would need to be upwards of 40% over the current levels of usage.

The preventative maintenance program, by the book, is not a financially viable project. However, because it only costs the company $847 over the course of 10 years, and has the potential to save $1,800 per year, it becomes a marginal project. The labor estimates for performing the maintenance work are conservative and most likely would be higher.

Unique Features of This Case

This case study reflects a cooperative company with relatively accessible and precise cost data. The manufacturing process is not complicated and lends itself to study as a solid example of how easy TCA can be.

CASE STUDY 6: Evaporator System

Step 1: Review of the Production Process

This small-size metal-finishing company applies coatings to a wide variety of parts for the automotive and consumer products industries. Prior to coating, these parts must be cleaned. Four process lines make up the cleaning department: zinc phosphating, iron phosphating, chrome conversion, and the stripping line. See Figure 7.4. This analysis was conducted to determine the financial impact of closing the loop on process waters.

The project will first minimize process water dragout by incorporating several pollution prevention techniques such as dragout tanks,

■ **FIGURE 7.4 Closed–Loop Rinse and Washwater System for Zinc and Iron Phosphating Lines**

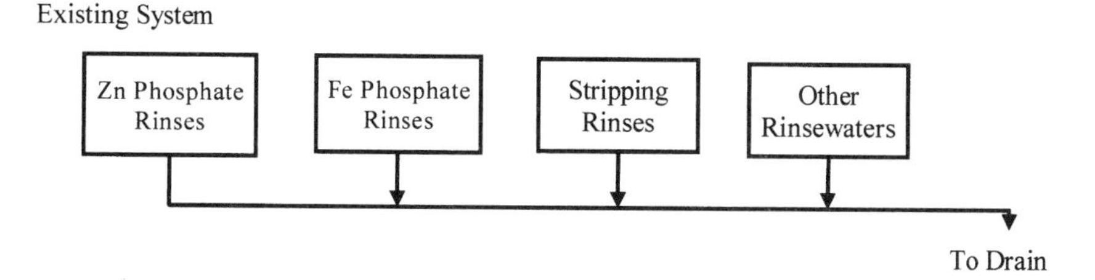

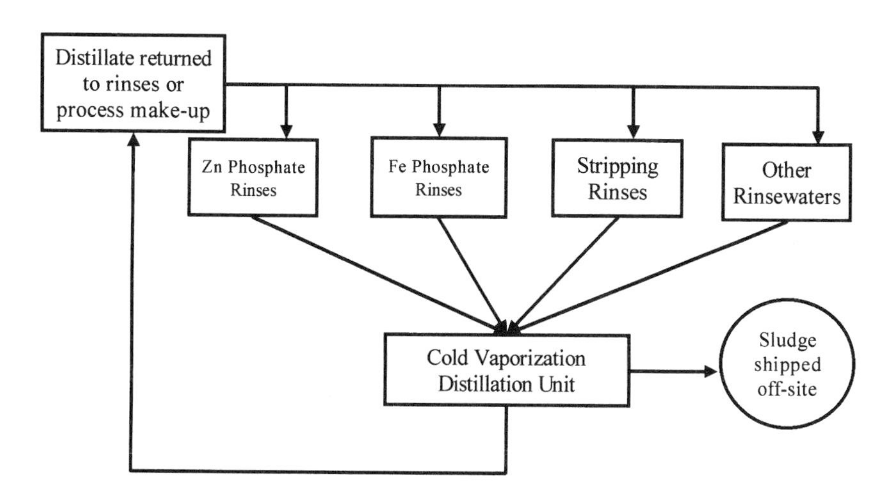

countercurrent rinses, enlarged drain holes in the parts barrels and baskets, and spray rinses to minimize the creation of wastes and rinsewaters. The rinsewaters that are generated will be piped to a closed-loop cold vaporization distillation unit that will condense rinsewaters. These condensed rinsewaters then will be returned to the rinse tanks or used as process water makeup.

Additional waters collected from other process lines (clean and precoat in the E-coat, powder coating, and liquid coating) also will be piped to the cold vaporization distillation unit, processed, and returned for reuse.

Step 2: Gathering Cost Data

The company needed a reliable estimate of the capital cost for this project. As a result, more time was spent developing capital cost data than operating cost data. Table 7.7 details the capital costs for this project.

Step 3: Defining Alternatives

Alternative 1: Cold Vaporization Distillation Cold vaporization distillation was chosen because of its track record in other industries and the flexibility to handle diverse process solutions. These solutions include rinses and wash waters from eight water-intensive processes: zinc phosphating, iron phosphating, stripping lines, and five other precoating steps in the powder and electrostatic (E-coat) process lines. The chrome conversion line and stripping line rinses will be handled separately to segregate the nonhazardous waste streams (spent zinc, iron phosphate, and stripping solutions) from the hazardous waste streams and reduce the overall volume of hazardous waste being generated by 70%.

Step 4: Conducting a Balanced TCA

The potential costs and savings from implementing these projects are detailed in Table 7.8. It is expected that the additional capital savings from the reduction in use of water and waste disposal will be offset by the increased electrical usage of the two evaporators. The reduction in potential environmental liability becomes a significant factor in the financial analysis even though it was not measured for this project.

TABLE 7.7 Itemized Costs for the Closed-Loop System

Cost Item	Unit Cost	Quantity or Hours	Subtotal Cost	Total Cost
Equipment				
Spray rinse parts & equipment	—	—	$15,120	
Spray rinse-tanks	$ 1,000	(2)	$ 2,000	
Wiring for spray rinse	$ 1/ft	1,800 ft	$ 1,800	
1800-gal holding tanks	$ 1,390	(2)	$ 2,790	
800-gal holding tanks	$ 690	(1)	$ 690	
PTU-600 evaporator	$68,648/hr avg	(1)	$68,648	
PTU-300 evaporator	$35,621	(1)	$35,321	
Sumps, pumps & piping	—	—	$21,324	
Equipment subtotal				$ 147,993
Labor				
Plans & specs	$ 74/hr avg	106 hr	$ 7,860	
Welder	$ 30/hr	32 hr	$ 960	
Plumber (spray rinse inst.)	$ 30/hr	(3) @ 48 hr	$ 4,320	
Electrician	$ 30/hr	(2) @ 32 hr	$ 1,920	
Permitting (consultant)	$ 62/hr avg	66 hr	$ 4,120	
Laboratory testing	—	—	$ 3,000	
Installation supervision	$ 65/hr	60 hr	$ 3,900	
Plumber (evaporator)	$ 30/hr	120 hr	$ 3,600	
Electrician (evaporator)	$ 30/hr	(2) @ 32 hr	$ 1,920	
Labor total				$ 31,600
Project subtotal				**$179,533**
20% contingency				$ 35,906
Total closed-loop costs				**$215,439**

■ TABLE 7.8 Financial Analysis of the P2 Project

Capital Costs	Description	Current Process	Closed Loop
Equipment purchase			$ 111,549
Disposal of old process			N/A
Research & design			$ 7,860
Initial permits			$ 4,120
Building/process changes			$ 36,444
Total capital costs			$ 159,973
Operating Costs	**Description**	**Current Process**	**Closed Loop**
Purchase costs	Chemicals		
	NaOH	$ 270	$ 216
	Muriatic acid	$ 119	$ 96
Storage/floor space	Chemical storage	$ —	$ —
	By-product storage	$ —	$ —
	Emissions storage	$ —	$ —
Waste management	Treatment chemicals	N/A	N/A
	Testing	$ 500	$ 250
	Disposal	$ 1,300	$ 650
Regulatory compliance	Safety/training equipment	Same	Same
	Fees or taxes		
	Base chemical fee	$ —	$ —
	Chemical fee	$ —	$ —
	Manifests, test reports	$ 40	$ 20
Insurance		N/A	N/A
Production costs	$/Unit of product	$ —	$ —
	Labor/unit product	$ —	$ —

(Continued)

■ **TABLE 7.8** *(Continued)*

Operating Costs	Description	Current Process	Closed Loop
Maintenance	Time	$ 3,000	$ 1,000
	Materials	$ 500	$ —
Utilities	Water	$ 4,350	$ 210
	Electricity	Same	Same
	Gas/steam	N/A	N/A
Annual operating costs		10,079	2,442
Cash Flow Summary	**Description**	**Current Process**	**Closed Loop**
Total operating costs		$ (10,079)	$ (2,442)
Incremental cash flow			$ 7,637
−Depreciation	10 years		$ (15,997)
Taxable income			$ (8,360)
Income tax (40%)			$ —
Net income			$ (8,360)
+Depreciation			$ 15,997
After-tax cash flow			$ 7,637
Present value	10 yr @ 10% = 6.1446		$ 46,926
Total capital cost			$ (159,973)
Net present value			$ (113,047)

Cost Assumptions

1. The equipment purchase costs of this project include all tanks and two evaporators. Research and design costs include the engineering design and specification costs of the project. Costs for the initial permits include consulting fees for filing a revised wastewater discharge permit. Building and process change costs include all wiring and process piping required to install the new equipment. Contractor labor (i.e., electricians, and plumbers) was not included in this analysis as a depreciable expense.
2. The ability to reuse the rinsewaters will significantly decrease the costs associated with supplying deionized water to several powder

and electrostatic coating lines. The decreased usage of citywater would require less use of the deionization columns and consequently less need to regenerate the columns. This reduced rate of regeneration decreases chemical usage. Estimates for this project place the usage reduction of acid and caustic at 20% and could be as high as 40%. Current costs for sodium hydroxide are $135/drum. Average annual usage is 1400 lb (2 drums/year), and costs the company $270. Muriatic acid currently sells for $.085/lb. The company used ten 140-lb Del drums last year, at a cost of $119.

A 20% cost savings through reduced chemical usage and chemicals present on-site would lower annual costs for NaOH and muriatic acid to $216 and $96, respectively.

3. The current treatment system for rinsewaters consists of diatomacious earth filters. These require a high degree of maintenance, roughly 200 hr/yr at $15/hr. This system will be eliminated with the closed-loop design. There will be some maintenance required for the evaporators: 64 hr/yr at $15/hr.
4. Industrial water usage amounts to $4,305/yr for both buildings. This represents over 1,805,400 gal/yr of water. The company expects a 95% reduction in water usage as a result of this project. This results in a decrease of over 1,715,000 gal/yr of water going to the sewer. The direct cost savings connected to this decrease is $4,305 × 95% = $4,089. Expected new water usage costs are $210/yr for industrial water use.
5. For the purposes of this analysis, a 10-year equipment lifetime was assumed. To keep calculations simple, the straight-line method of depreciation was used. The company's cost of capital is 10% and a corporate tax rate of 40% was also assumed.

Financial Analysis Conclusions

Given the data available, this project would not benefit the company financially in the short term. The net present after-tax value is ($113,047). This means it will cost the company this amount over the life of the project. The payback for this project is over 21 years, even though it saves the company $7,637/yr. The small quantities of chemicals used contribute to the low operating costs and resultant poor financial return. The intangible benefits include reduced worker exposure to chemicals, reduced depletion of local water supplies, and reduced regulatory liability from having a sewer discharge and large quantities of wastes.

Unique Features of This Case

In this case, it would make sense to proceed and attempt to quantify the contingent costs. Also note the lack of production data. Other closed-loop metal-finishing companies have reported enormous benefits form being closed-looped. One plant in particular was able to continue operating its shop even when the water main in the industrial park broke. Then when the main was restored, most of the water users received rust-loaded water that forced some to dump their process baths and start from scratch. The closed-loop company had no such difficulties. What is that worth in terms of production costs?

CASE STUDY 7: Jewelry Cleaning

This nationally recognized manufacturer specializes in name plates and badges for federal, state, and local law enforcement departments. Customer satisfaction hinges largely on appearance and other aesthetic qualities. Because of this, great care is taken in cleaning the product and minimizing scratches and surface blemishes. The company desired to find a solvent substitute that would reduce environmental liability yet maintain the high standard of product appearance. This TCA was performed after the fact to highlight the benefits of switching away from chlorinated solvents.

Step 1: Review of the Production Process

The manufacturing processes include stamping of copper stock, deburring, cleaning, plating, engraving, enameling, and buffing. The plant manager's previous experience in solvent substitution at a similar facility helped him select several possible alternatives. The company chose an aqueous system to avoid potential regulatory or disposal issues associated with some of the other cleaners. Figure 7.5 shows the cleaning process before and after the project.

Step 2: Gathering Cost Data

The company has a computer database of materials in inventory that facilitated gathering data for this case study. The plant's environmental

■ **FIGURE 7.5 Jewelery Degreasing System**

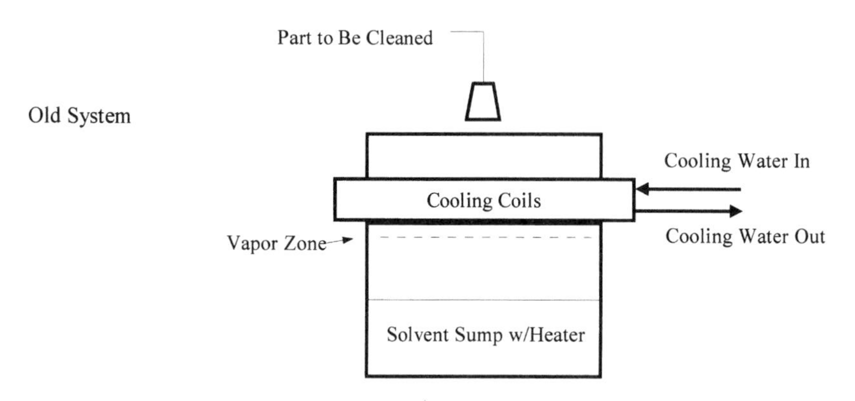

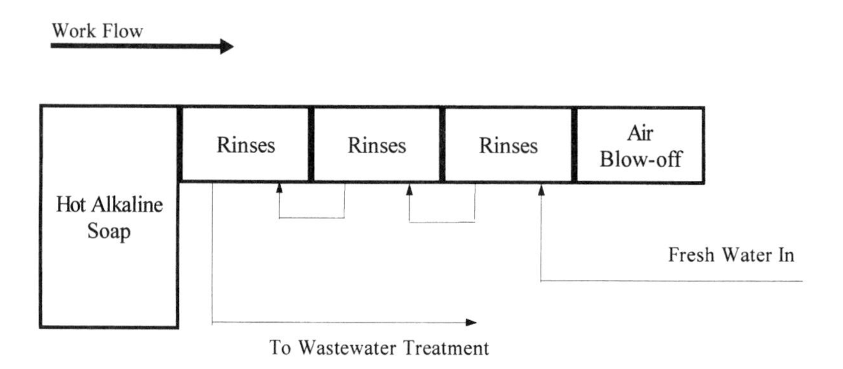

and electroplating manager was able to estimate costs not found in the computer system. Operating costs for the old system are shown in Table 7.9.

1. The existing vapor degreaser was 10 years old when replaced; no salvage value was assigned to it.
2. The company's 1989 usage of trichloroethylene (TCE) was 25,492 pounds. TCE sold for $0.48/lb in 1989, or $5.80/gal. At the time of the analysis, the market price for TCE was $1.08/lb, or $12.50/gal.[1] If the company had continued using TCE, it would be spending $27,531/yr on purchases.

[1]Ashland Chemical Company market price quote, December 1993.

TABLE 7.9 TCA of New Cleaning Process

Capital Costs	Description	Old Process	New Process
Equipment purchase			$ 9,000
Disposal of old process			$ (48)
Research & design			$ —
Initial permits			$ —
Building/process changes			$ 432
Total capital costs			$ 9,384
Operating Costs	**Description**	**Old Process**	**New Process**
Purchase costs	Chemicals	$27,531	$ 444
Storage/floor space	Chemical storage	$ —	$ —
	By-product storage	$ —	$ —
	Emissions storage	$ —	$ —
Waste management	Treatment chemicals	$ —	$ —
	Testing	$ —	$ —
	Disposal	$ 1,320	$ —
Regulatory compliance	Safety/training equipment	$ 20	$ 20
	Fees or taxes	$ 1,100	$ —
	Manifests, Test Reports	$ 240	$ —
Insurance		N/A	N/A

(Continued)

■ **TABLE 7.9** ***(Continued)***

Production costs	$/unit of product	$ —	$ —
	Labor/unit product	$ —	$ —
Maintenance	Time	$ 112	$ 576
	Materials	$ —	$ —
Utilities	Water	N/A	$ 7
	Electricity	Same	Same
	Gas/steam	Same	Same
Annual operating costs		30,323	$ (1,047)
Cash Flow Summary	**Description**	**Old Process**	**New Process**
Total operating costs		$(30,323)	(1,047)
Incremental cash flow			$ 29,276
−Depreciation			$ (938)
Taxable income			$ 28,338
Income tax (40%)			$ (11,335)
Net income			$ 17,003
+Depreciation			$ 938
After-tax cash flow			$ 17,941
Present value	10 yr @ 10% = 6.1446		$110,240
Total capital cost			$ (9,384)
Net present value			$100,856

3. Disposal of waste TCE depends on the percent TCE per drum and runs between $165 and $325/drum. Current annual disposal costs would be $1,320/yr for eight drums.[2] This is up from the 1989 level of $880/yr for the same method of disposal—solvent reclamation.
4. Usage levels of TCE required the company to file both SARA 313 form R and TURA form S. The TCE use fee under the state pollution prevention law was $1100/yr. In addition to the fee, the associated paperwork took 15 hr for both forms and another 3 to 4 hr for permits associated with the chemical use.
5. Maintenance on the old system was significantly less than the new unit, requiring a clean out once every 6 months (7 hr/yr).

Step 3: Defining Alternatives

As mentioned earlier, the company had made the switch to a nonchlorinated cleaning process when the TCA was performed. The new process consists of an aqueous detergent, air agitation, and countercurrent rinsing. The unit was custom-built for $7,200 in 1991. This custom system has since become a product line for the contractor and sells for $9,000.[3] No additional changes to the processes or building were necessary; however, labor associated with disconnecting the old process and connecting a water line for the new system has been included. The old vapor degreaser was cleaned out and sold as scrap for $48.

Step 4: Conducting a Balanced TCA

The TCA was used to corroborate the value of pollution prevention and build support for future projects within the company. The data on the new process was more readily available than the old process and did not require adjustments because the costs were already in current dollars.

Cost Assumptions

1. The new system uses an alkaline aqueous chemistry. Chemical usage for the system averages 0.5 gal/wk of makeup. This is significantly

[2]East Coast Environmental Services price quotes, December 1993.

[3]Price figure from an interview of the Greco sales staff, December 1993.

less usage than the 10 to 20 gal/wk of TCE previously used. The cost per gallon for the new chemical is slightly higher than TCE's current market price ($12.95/gal for new cleaner versus $12.50/gal for TCE). The aqueous chemicals last longer and have not yet been disposed.

2. The aqueous system does not trigger any regulatory thresholds. The company reduces the number of environmental reports and manifests it must file.
3. The company had not consulted with its insurance agent at the time of the interview, but was interested in pursuing premium decreases due to fewer toxics being on-site. The company is saving $1,400/yr on avoided paperwork and fees, yet it has maintained regulatory compliance by eliminating regulated substances.
4. Production costs have not increased as a result of this project. Original operator practices were to leave racks or baskets of parts hanging in the vapor degreaser and do another task. The new system allows them to operate in the same way.
5. Maintenance on the new system is higher due to more frequent filter changes and cleanouts (monthly cleaning on the new unit, 36 hr/yr). Copper filings are cleaned from the bottom of the tank every month and sold to the local scrap dealer. The in-line filters are changed every 6 months and discarded as solid waste.
6. The running rinses consume 60 gal/day of water, 15,000 gal/year, or roughly $7/yr. Compared to water used in the plating and grinding areas, this is a nominal impact. The company used in-plant steam to heat both the old and new systems. Pump size and horsepower were also similar between the two parts cleaners. Because both systems were/are run for a single 8-hr shift, the cost difference for steam heat was considered marginal.

Financial Analysis Conclusions

Current annual cost for running the old degreaser would be $30,323. The total annual savings from the new system is $17,972 after taxes. The net present value on this project is $100,856. This means that the investment in an alternative to TCE is returning over $10 for every dollar invested.

Notice that the old degreaser was disposed of at a profit. The company assisted in the cleanout of the system, and the scrap dealer furnished the transportation and labor to remove the unit. Copper filings from the cleaned parts accumulate at the bottom of the new cleaning tank. These filings are also sold as scrap, but are thrown together with other copper scrap from the stamping operation. The price for this waste stream could not be estimated by plant personnel.

Unique Features of This Case

One advantage of a retrospective TCA is the quality of postdecision data. Typically, the data used for the "alternative" are more an average or expected cost than a number pulled from accounting records. Looking back allows you to see how well your decisions turned out. Some people call this process a "project autopsy" and is commonly performed in the consulting industry to develop more accurate project budgets.

CASE STUDY 8: Machine Tool Coolants

A medium-size precision metal working company machines aluminum, titanium, stainless steel, and brass in manufacturing aerospace and defense contractor products. Having found a unique solution to a common problem of metalworking coolant spoilage, the company wants to know if it is financially feasible to install the new technology on other metalworking machines sumps. See Figure 7.6.

■ FIGURE 7.6 Process Flow Diagram for Metalworking Coolants

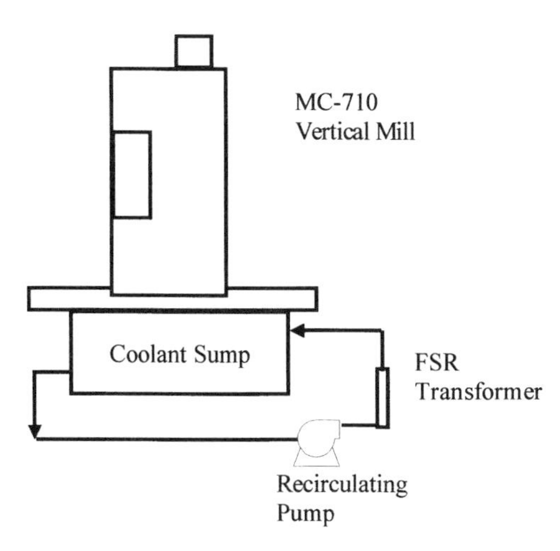

Step 1: Review of the Production Process

The variety of metals machined and infrequent use lend themselves to bacterial growth. The machine, a Mitsura MC-710 vertical mill with a 15-gal sump, characterized the typical Monday morning odor and operational problems that commonly affect metalworking lubricants. Despite repeated efforts to correct the problems, the odors became so bad that the operator refused to work at the machine.

Biocides were applied and resulted in operator dermatitis. Even with the addition of biocides, the odors persisted, and eventually the bacteria acclimated to the new biocide. Every 3 weeks, the machine sump was emptied and flushed with a machine cleaner. This resulted in increased waste disposal costs and liabilities for the company. The plant environmental manager recognized an opportunity for a fresh approach to the problem and installed an "electronic biocide" device to control pH and bacterial growth.

This electronic biocide device continuously circulated the fluid through its forced sequential rephasing transformer. This device solved the odor issue. Monthly testing of the coolant bath revealed significantly lowered levels of bacteria, no spoilage, and no odor. The same coolant has been in use for 18 months. Most importantly, the machine still holds its tolerance of 32-micron finishes and eliminated staining of brass and aluminum parts.

Step 2: Gathering Cost Data

Interviews with the plant environmental manager provided almost all the cost data required for this analysis. Additional information was provided by the accounting department and purchasing records. Laboratory analysis of bacteria counts was used to determine the number and type of bacteria present in the machine sump after installation of the unit. The items that follow explain the costs shown in Table 7.10 under the "Current Process" heading.

Cost Assumptions

1. Coolant changeout occurs every 3 weeks with the traditional system. This changeout uses 14 gal of water and 1 gal of coolant concentrate having an average cost of $6.00/gal.
2. Flushing the machine with cleaner occurs at the same time and with the same frequency as the coolant changeout. It would be a waste of

TABLE 7.10 TCA of Electric Biocide Technology

Capital Costs	Description	Current Process	Electric Biocide
Equipment purchase			$ 5,000
Disposal of old process			$ —
Research & design			$ —
Initial permits			$ —
Building/process changes			$ —
Total capital costs			$ 5,000
Operating Costs	**Description**	**Current Process**	**Electric Biocide**
Purchase costs	Chemicals		
	Coolant concretrate	$ 144	$ 2
Storage/floor space	Chemical storage	$ —	$ —
	By-product storage	$ —	$ —
	Emissions storage	$ —	$ —
Waste management	Treatment chemicals	$ —	$ —
	Testing	$ 50	$ —
	Disposal	$ 993	$ —
Regulatory compliance	Safety/training equipment	$ —	$ —
	Fees or taxes	$ —	$ —
	Base chemical fee	$ —	$ —
	Chemical fee	$ —	$ —
	Manifests, test reports	$ 180	$ —
Insurance		N/A	N/A

(Continued)

TABLE 7.10 *(Continued)*

Production costs	(Form down time)	$ 1,920	$ —
	Labor/unit product		
Maintenance	Time	$ 480	$ 47
	Materials	$ 720	$ 10
Utilities	Water	$ 1	$ —
	Electricity	$ —	$ 20
	Gas/steam	N/A	
Annual operating costs		4,488	79
Cash Flow Summary	**Description**	**Current Process**	**Electric Biocide**
Total operating costs		$(4,488)	$ (79)
Incremental cash flow			$ 4,409
− Depreciation			$ (500)
Taxable income			$ 3,909
Income tax (40%)			$ (1,564)
Net income			$ 2,345
+ Depreciation			$ 500
After-tax cash flow			$ 2,845
Present value	10 yr @ 10% = 6.1446		$ 17,484
Total capital cost			$ (5,000)
Net present value			$ 12,484

new coolant not to clean out the machine and remove possible bacteria breeding sites. Of the 15 gal used in this maintenance step, 5 gal are pure cleaner concentrate, which sells for $6.00/gal.

3. Disposal of coolant and cleaner includes all fees and taxes associated with disposal of a state-classified hazardous waste. This machine generates 30 gal of waste coolant and cleaner every 3 weeks. The disposal cost for this waste is $1.38/gal. Time has been allocated for the manifesting and labeling drums of the waste coolant and for testing the coolant.
4. The machine is down for unscheduled maintenance every 3 weeks, which interrupts the production schedule and costs the company money. This machine is charged out at $80/hr and is down 1 hr every 3 weeks, or a full 24 hr/yr.
5. There is additional maintenance labor required for cleaning out the machine every 3 weeks. This takes one person 1 hr to complete. The average fully loaded labor rate is $20/hr.
6. Utility costs for the chemically maintained coolant sump were very low. Water use is only 800 gal/yr for this sump. Table 7.10 shows a cost of $1 even though it is probably less than that. The machine uses no electricity, steam, or natural gas to maintain the coolant. Circulation of the fluid through the machine while it operates would be the same under both conditions and is not included.[4]

Step 3: Defining Alternatives

In this case, the environmental manager already knew which alternative would be tried. He selected a forced sequential rephasing transformer and a small recirculating pump. Initially, just the pump was installed to see if additional movement would prevent odors. Once that was disproved, the electronic biocide device was installed. It has operated continuously for 16 months, with only makeup solutions added to the sump—there is no odor, no chemicals, just the cost of the electricity. An added side benefit is an improvement in product quality. Previously, some types of metals were stained by the rancid coolant and required polishing. There has been no staining since the installation of this technology.

[4]Cost data from an interview with F. Rogers, Piat Corp., Milford, Connecticut, December 1996.

Step 4: Conducting a Balanced TCA

Costs for the new system are minimal. We included a complete change-out and cleaning of the machine once every 3 years. This cost was annualized and appears in chemical purchases, maintenance labor, and materials. Water use has decreased and become even less significant. The only other major cost was electricity. The manufacturer of the technology claims it uses approximately 200 watts continuously, slightly more than a pair of light bulbs left on all the time. Table 7.10 shows the details of the TCA.

Financial Analysis Conclusions

The fact that the electronic biocide device eliminated a chronic problem satisfied the company. However, as company management contemplated, the question of how much money might be saved arose. The new system has a net present value of $12,484. It pays for itself in slightly more than 1 year and has an indefinite lifetime because there are no moving parts or sacrificial anodes to wear out. For the purposes of the analysis, a 10-year lifetime was assigned; however, it may operate well beyond 10 years. Estimated revenues for installing the "electric biocide" on all metalworking machines are approximately $126,000 annually.

Other benefits from this technology include the following:

- Elimination of the offensive odor, thus improving working conditions and worker morale
- Elimination of the staining of aluminum and brass parts when the coolant went rancid
- Reduction in volume of state-regulated wastes and associated liability

CASE STUDY 9: Food Vessel Cleaning

Step 1: Review of the Production Process

This national producer of juice and juice beverages mixes all ingredients in large stainless steel vats. Following product runs, these vessels must be cleaned to Food and Drug Administration (FDA) cleanliness specifications. A blend of phosphoric acid and nitric acid is used as a sanitizer that restores a protective chromium oxide coating on stainless steel surfaces. It is used when cleaning blend vessels, filters, pectin tanks, and connecting lines; feed tank lines to product filler, which include the pasteurizer, surge tanks, the filler itself; and return lines, juice storage tanks, as well as tank loading and off-loading lines. The chemicals are introduced manually or automatically to a vessel or connecting line so that all tanks, lines, and so forth, that previously held product become sanitized. Figure 7.7 shows the relationship of the cleaning equipment and fluids to the production process.

Step 2: Gathering Cost Data

The company was primarily concerned about using cleaning chemicals and desired to reduce or eliminate them wherever feasible. The data gathering goal was to determine how much the primary cleaning chemical (phosphoric acid) cost the company. Fortunately for us, the number

■ FIGURE 7.7 Process Flow Diagram for Food Vessel Cleaning

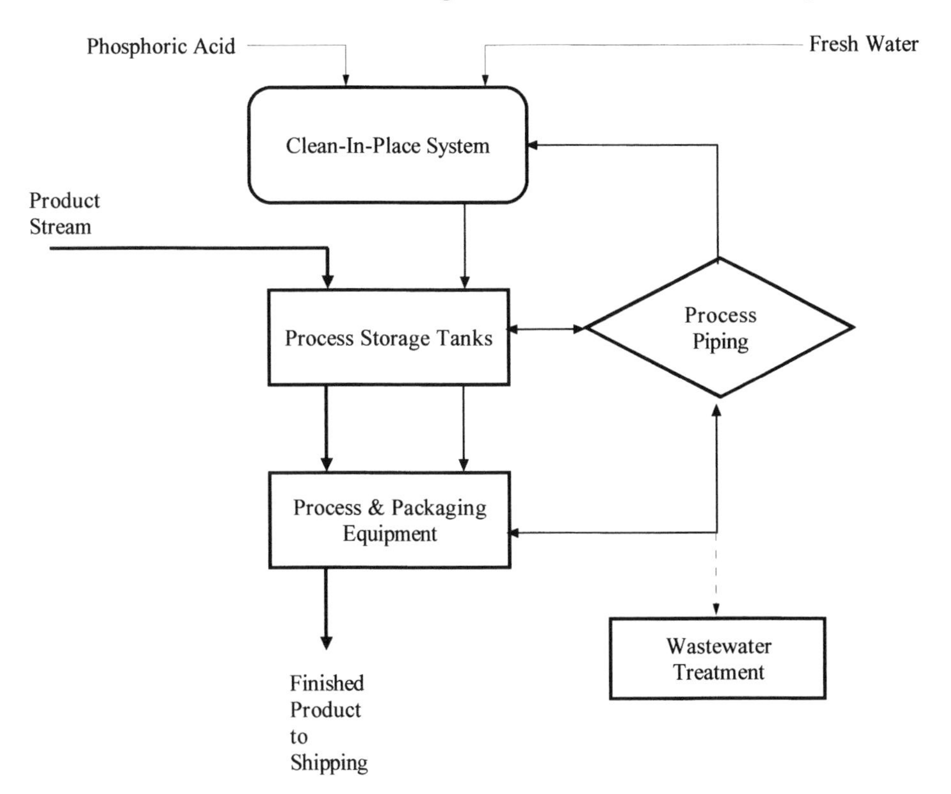

crunchers, this company had a computer inventory system for all chemicals that came on-site. Gathering cost data was simply a matter of discussing our needs with the environmental manager, who promptly pulled up all environmental costs on his computer. Other costs not directly charged to the environmental program were retrieved from the accounting systems as needed. Several cost categories did not have data either because they were not measured or did not apply. See Table 7.11 for specifics.

Cost Assumptions

1. The company was experiencing a surplus of warehouse space at the time and storage costs for chemicals were not calculated. The company owns the warehouse space and therefore would lose money only if there were a shortage of space, or if the space could "earn" more money by being occupied by production equipment. This is an example of the opportunity cost of storage space and its relative value to a firm.
2. Long-term liability connected with the use of phosphoric acid is considered to be low. It is a reactive substance, highly corrosive and hazardous to workers; it does not contaminate drinking water, cause mutations, or cancer, and does not deplete the ozone or contribute to global warming. The largest risk comes from possible fines for exceeding the pH limits of the local sewer authority and worker exposure to a dissolved solution of phosphoric acid. Because of this limited risk potential, insurance costs are not a factor in this analysis and the impact of public perception is also a low-weighted factor.
3. Waste management costs for this chemical are minimal because the quantities used to clean the production piping and tanks are small compared to the other wash waters containing more neutral and basic pH levels. The over all juice effluent tends toward the acidic side from the tanins in the fruit skins.
4. Production costs associated with this chemical are minimal. Cleaning is a necessary part of maintaining the equipment and has no direct bearing on the batch operations, because cleaning would be required between each batch to prevent flavor mixing.
5. Maintenance costs associated with a maintenance chemical is redundant. No costs exist for this category.

TABLE 7.11 Operating Cost for Using Phosphoric Acid

Cost Item	Description	Cost Factor	Unit Price	Total Units	Total Cost
Purchase costs	Chemicals Phosphoric acid	100 %	\$3.55/gal	2,481 gal	\$ 8,807
Storage/floor space	Ancillary storage	—	—	—	\$ —
	Chemical storage	N/A	N/A	N/A	N/A
	By-product (scrap) storage	—	—	—	N/A
	Emissions (waste) storage	—	—	—	N/A
Waste management	Treatment chemicals	Minimal	—	0	\$ 50
	Testing	100 %	—	—	\$ 320
	Disposal	50 %	\$ 4,250	—	\$ 2,120
Regulatory compliance	Safety/training equipment	25 %	\$ 11,180	—	\$ 2,795
	Fees & Taxes				
	Base fee	25 %	\$ 4,625	—	\$ 1,156
	Chemical	100 %	\$ 1,100	—	\$ 1,100
	Manifests, test reports	25 %	\$ 460	—	\$ 115
Insurance		—	—	—	
Production costs	\$/Units of product	—	—	—	N/A
	Labor/Unit product	—	—	—	N/A
	Other	—	—	—	N/A
Maintenance	Time	—	—	—	N/A
	Materials	—	—	—	N/A
Utilities	Water	50 %	\$3.4/1000 ft^3	3.75 mm	\$ 6,375
	Electricity	—	—	—	N/A
	Gas/steam	—	—	—	N/A
Total annual operating cost					\$22,838

6. There is an annual sewer usage charge of $4,125. This was divided by the three chemicals most used for cleaning. This best approximates the impact of phosphoric on pretreatment.
7. Water costs have also been proportioned based on phosphoric acid usage in relation to the other cleaning chemicals used.

Step 3: Defining Alternatives

Alternative 1: Automating the Cleaning Process The juice storage tank farm must be cleaned after each batch flavor change. These tanks are currently cleaned by flooding and agitating, using hoses. The use of a high-efficiency spray washer and spray ball will minimize water use and also chemical use. Installation of spray ball cleaning could result in an estimated 50% reduction in water and chemical use in this area. The company also estimates a reduction in cleaning time and labor required to clean the tanks.

During installation of the spray baths, the pipes will be redesigned to eliminate bends, joints, overpasses, and to take advantage of gravity in delivery of product. The company anticipates eliminating large sections of pipe needing cleaning, reducing the need to clean by 5–10%. A financial analysis of this option follows.

Step 4: Conducting a Balanced TCA

The company was eager to see if the estimated reductions would result in a profitable venture. Conservative numbers were applied to the TCA to give a worst-case scenario. The following conditions were applied to the evaluation of the project. Table 7.12 shows the full TCA.

Cost Assumptions

1. Purchase costs for the spray balls total $1,200 to retrofit the whole tank farm. Vendor prices and invoices for projects conducted elsewhere in the facility provided these figures. Building and process changes account for the largest cost of this project. Total estimates are over $80,000 for repiping and installing new systems. One-third of this cost is allocated to the use of phosphoric acid in keeping with past allocations based on the three cleaning chemicals used.
2. Phosphoric acid usage was below the reporting threshold of 10,000 lb. However, production increases are expected and this usage may

■ **TABLE 7.12 Financial Analysis of the P2 Project**

Capital Costs	Description	Current Process	Auto Clean
Equipment purchase			$ 400
Disposal of old process			
Research & design			
Initial permits			
Building/process changes			$26,933
Total capital costs			$27,333
Operating Costs	**Description**	**Current Process**	**Auto Clean**
Purchase costs	Chemicals	$8,807	$7,050
	—	—	—
Storage/floor space	Chemical storage	N/A	N/A
	By-product storage	N/A	N/A
	Emissions storage	N/A	N/A
Waste management	Treatment chemicals	$ 50	$ 25
	Testing	$ 320	$ 320
	Disposal	$2,120	$ 1,060
Regulatory compliance	Safety/training equipment	$2,795	$ 2,795
	Fees or Taxes		
	Base chemical fee	$1,155	$ —
	Chemical fee	$1,100	$ —
	Manifests, test reports	$ 115	$ 115
Insurance		N/A	N/A

(Continued)

■ TABLE 7.12 ***(Continued)***

Capital Costs	Description	Current Process	Auto Clean
Production costs	$/Unit of product	N/A	N/A
	Labor/unit product	N/A	N/A
Maintenance	Time	$ —	$ —
	Materials	$ —	$ —
Utilities	Water	$ 6,375	$ 3,187
	Electricity	N/A	N/A
	Gas/steam	$ —	$ —
Annual operating costs		22,837	14,552
Cash Flow Summary	**Description**	**Current Process**	**Air Purge**
Total operating costs		$(22,837)	$(14,552)
Incremental cash flow			$ 8,285
−Depreciation			$ (2,733)
Taxable income			$ 5,552
Income tax (40%)			$ 2,221
Net income			$ 3,331
+Depreciation			$ 2,733
After-tax cash flow			$ 6,064
Present value	10 yr @ 10% = 6.1446		$ 37,263
Total capital cost			$(27,333)
Net present value			$ 9,930

rise in following years. The zeros in this category show that despite increased production, this project will reduce quantities below reportable thresholds.

Financial Analysis Conclusions

Because all the costs and cost savings were allocated based on the three chemicals used in this process, this analysis shows only one-third of the project's expected costs and savings. The other savings will depend on chemical usage reductions and unit costs of the other chemicals. This portion of the project has an overall net present value of $9,928. The return on investment is 20%, which is higher than the assumed cost of capital of 10%. Therefore, this project is financially feasible.

Unique Features of This Case

It is important to note that maintenance activities such as cleaning juice vessels do not usually require maintenance in themselves. There may be instances where gauges used to check process equipment have to be calibrated or scales are balanced, in which case maintaining the equipment that performs the maintenance may be an appropriate cost. Typically, manufacturing plants schedule down time for maintenance. This time would not be necessarily accounted for unless a pollution prevention project reduces or eliminates the need for the maintenance down time. Unexpected down time or chronic breakdowns may be worth considering if it is known the alternative will have a superior level of reliability.

CASE STUDY 10: Architectural Hardware

This medium-size company manufacturers replicas of antique architectural hardware. These parts are stamped from brass, copper, and steel stock and plated or painted to resemble original decorative trim, strike plates, vent grills, ceiling and wall fixture plates, and other hardware. In order to work metal stock of this thickness, the parts must be annealed prior to striking. See Figure 7.8. The company would like to find ways to optimize the amount of product sent through the annealing furnace given that it is more economical to keep the furnace on all the time.

Step 1: Review of the Production Process

The company uses a conveyorized annealing furnace to heat treat the parts. Anhydrous ammonia is cracked to provide an oxygen-free atmosphere without changing the color of the metal. The parts are scheduled to pass through the furnace and sometimes require more than one pass if the design is not impressed deep enough or if more than one design is being applied. Following the stamping and embossing operations, the products flow to plating, polishing, and inspection stations.

■ **FIGURE 7.8 Process Flow Diagram of Annealing Furnaces**

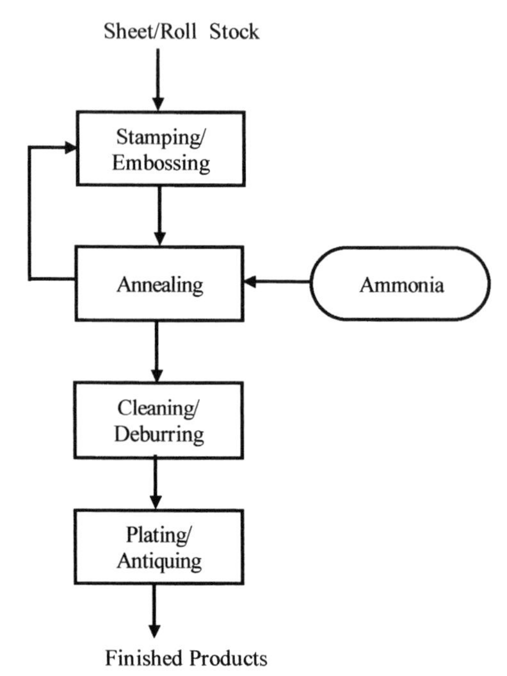

Step 2: Gathering Cost Data

To determine if improvements are financially feasible, the operating cost of the furnace will be compared to the capital cost of proposed changes. The company uses an index card form of materials tracking for some chemicals and relies on paper records for tracking production and scheduling parts flow. This makes obtaining cost data rather tedious if the information requests are numerous or involve several years' worth of records. Table 7.13 shows the operating costs for annealing.

Cost Assumptions

1. Ammonia is the only chemical input for this process. It is assumed to be completely combusted once inside the furnace. Leaks in the delivery system would be detected by monitors or employees.
2. There are regulatory reporting and planning requirements associated with the use of this chemical. The company uses several chemicals in

■ **TABLE 7.13 Operating Cost for Annealing**

Cost Item	Description	Cost Factor	Unit Price	Total Units	Total Cost
Purchase costs	Chemicals Ammonia	100%	$ 0.32	22,616	$ 7,237
Storage/floor space	Chemical				N/A
Waste management	Treatment chemicals	0%			$ —
	Testing	0%			$ —
	Disposal	0%			$ —
Regulatory compliance	Safety/training equipment	100%	$ 140	1	$ 140
	Fees & taxes				N/A
	Manifests, test reports	100%	$ 32	43	$ 1,376
Insurance					N/A
Production costs	$/product				$ —
	Hours/product				$ —
	Outsourced work	100%	$13,310	1	$13,310
Maintenance	Time	100%	$ 25	215	$ 5,375
	Materials	100%	$ 1,500	1	$ 1,500
Utilities	Water	35%	$ 3.15	4,250	$ 4,686
	Electricity	12%	$ 0.07	1.65E+06	$14,078
	Gas/steam				N/A
Total annual operating cost					$47,702

quantities above the SARA 313 thresholds and generates a large quantity of waste. The ammonia does not contribute to the water or hazardous waste costs of the company.

3. Long-term liability connected with the use of ammonia is low. The largest risk during an ammonia leak comes from damage to the mucus membranes of the eyes, nose, lungs, and throat. Because of this limited risk, potential insurance costs are not a factor in this analysis, and the impact of public perception is also a low-weighted factor.

Step 3: Defining Alternatives

To meet the goal of optimizing material flow through the annealing furnace, the constraints of the system must be examined. The furnace is needed to provide a softer strike surface, allowing the metal to be worked without cracking, and to create a sufficiently deep imprint. The constraints are the number of pieces, or repeats of the same piece, that pass through the furnace, and the rate at which they pass through. Minimizing the number of trips a piece makes through the furnace could provide a capacity increase. Increasing the temperature and the travel rate of the piece also could increase the throughput. The furnace is already operating at its maximum specified temperature. Exceeding this temperature could cause degradation of seals and conveyor parts.

The Alternative: Tighter Process Control Better control of how the parts are stamped, the pressure delivered by the press, and when the parts are sent to the annealing furnace could influence the capacity of the system. The craftsman operating the presses have many years of experience but no written directions on how to set up the presses. As a result, other people working the presses often overwork the piece, resulting in too many trips through the furnace.

This project will create a system for identifying the part number, its material requirements, process steps, average length of time for each process step, and average number of labor hours required. Included in this description will be the press settings, setup instructions, and other information to help reduce overwork of parts.

Part of this effort will also involve scheduling improvements to eliminate last minute rush orders. Eliminating the bottleneck at the furnace would allow shorter lead times and better scheduling.

Step 4: Conducting a Balanced TCA

Designing and implementing a new process control system requires a large capital investment, but very little operating expense. Table 7.14 shows the estimated costs for this project.

Cost Assumptions

1. Research and development of the process control project will require at least two people full time for 3 months at $19/hr.

■ **TABLE 7.14 Financial Analysis of the P2 Project**

Capital Costs	Description	Current Furnace	Process Control
Equipment purchase			$ —
Disposal of old process			$ —
Research & design			$19,000
Initial permits			$ —
Building/process changes			$ 2,000
Total capital costs			$21,000
Operating Costs	**Description**	**Current Furnace**	**Process Control**
Purchase costs	Chemicals Ammonia	$ 7,237	$ 7,237
Waste management	Treatment chemicals	$ —	$ —
	Testing	$ —	$ —
	Disposal	$ —	$ —
Regulatory compliance	Safety/training equipment	$ 140	$ 140
	Fees or taxes	$ —	$ —
	Manifests, test reports	$ 1,376	$ 1,376
Insurance		N/A	N/A
Production costs	$/product	$ —	$ —
	Hours/product	$ —	$ —
	Outsourced work	$13,310	$ 3,000
Maintenance	Time	$ 5,375	$ 5,376
	Materials	$ 1,500	$ 1,500
Utilities	Water	$ 4,686	$ 4,686
	Electricity	$14,078	$14,078
	Gas/steam	N/A	N/A
Annual operating costs		47,702	37,393

(Continued)

■ TABLE 7.14 *(Continued)*

Cash Flow Summary	Description	Current Furnace	Process Control
Total operating costs		$ (47,702)	$(37,393)
Incremental cash flow			$ 10,309
−Depreciation			$ (2,100)
Taxable income			$ 8,209
Income tax (40%)			$ (3,284)
Net income			$ 11,493
+Depreciation			$ 2,100
After-tax cash flow			$ 13,593
Present value	10 yr @ 10% = 6.1446		$ 83,521
Total capital cost			$(21,000)
Net present value			$ 62,521

2. Consider the costs of retraining the workers as part of the building and process change cost category.
3. The company projects a 30% increase in the throughput of the furnace due to reductions in overwork and better setup.
4. The company outsources some work when scheduling becomes impossible. In the future, the company will only outsource rush jobs, as these tend to have higher value and are the largest scheduling problem. The cost savings shown in Table 7.14 results from a reduction in the total quantity of parts sent out for annealing.
5. It is not anticipated that this project will directly decrease the use of any of the utilities. The furnaces will remain on all the time, but they will be more efficiently used.
6. There is no anticipated decrease in maintenance costs with the new process control system.

Financial Analysis Conclusions

The project has a positive net present value of $62,000 over 10 years. Implementing the process control improvements will result in at least $10,000 of cost savings per year.

Unique Features of This Case

It is important to note that most of the impact from this project comes through more efficient use of existing production capacity. Although there is little change in pollution output or chemical use at a facility level, the per piece usage of ammonia and natural gas has decreased. This points out some of the subtleties inherent in TCAs. Many times, the benefits are not plainly obvious, yet the projects are very profitable.

CASE STUDY 11: Textile Dyes

A division of a multinational textile corporation has had repeated difficulty meeting discharge limits for particular metals. The overdischarges are numerous enough that the company now faces legal action by the local sewer authority. The company president desires to resolve this matter as quickly and economically as possible. The pollution prevention team has conducted a review of the processes and has MSDS for all chemicals entering the facility. We are to negotiate with the sewer authority, set an implementation schedule, and determine the financial merits of the alternatives developed by the pollution prevention team.

Step 1: Review of the Production Process

The company purchases large bolts of fabric from all over the world, desizes, scours, dyes, and ships them to sister plants for use in garment manufacturing. The process flow is remarkably similar to an electroplating line both in layout and chemical makeup. Each has precleaning process steps and each has metals in solution of water and other chemicals, with the goal of transferring the metals and chemicals onto products. The bolts of fabric are then rinsed, stretched, and dried. The baths are heated with steam from the boilers and boiler blowdown is discharged to the drain. See Figure 7.9.

The rinsewaters flow to common lines and are discharged to the sewer with only pH adjustment. Although this may seem inadequate for the 800,000 gal/day of water handled, it appears only certain production runs result in effluent exceeding permitted limits.

All the dyes and additives are suspect for contribution to the effluent overages. However, testing the rinses does not result in sufficient levels of metals to explain the noncompliant readings. At first, this puzzled the pollution prevention team until samples of the imported cloth

■ **FIGURE 7.9 Process Flow Diagram for Textile Dyeing**

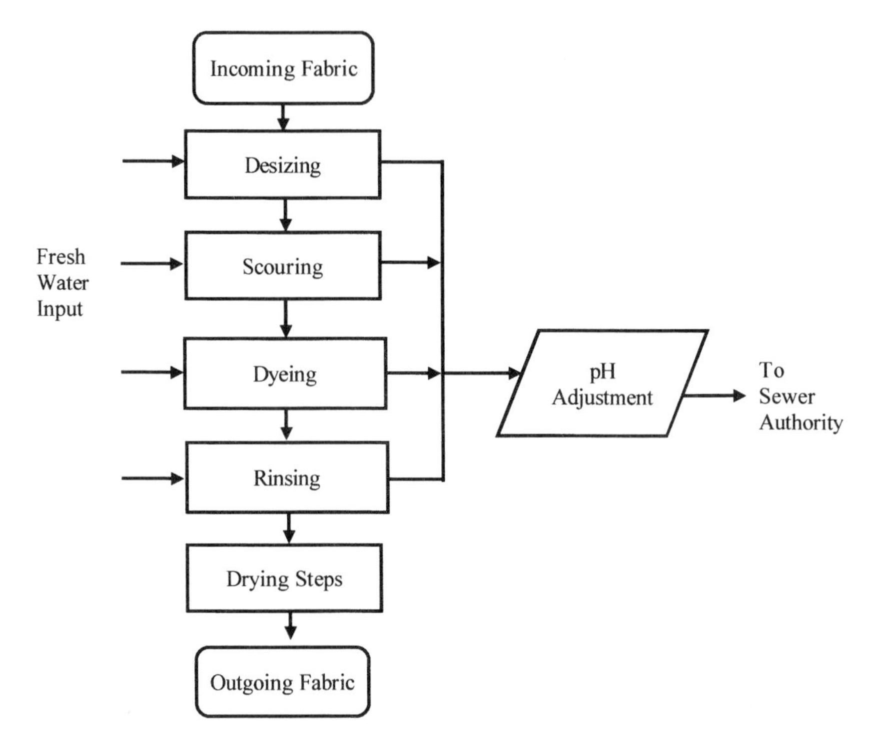

were examined. One of the team members found large amounts of the troublesome metal on the incoming fabric. Apparently, metal-containing compounds are used as rodenticides and fungicides on the fabrics' journey by boat to the United States.

Having found the culprit of the noncompliant readings, the team developed alternatives to achieve compliance. Our job as evaluators was to gather the data necessary to determine which option would cost the least amount of capital.

Step 2: Gathering Cost Data

The company had a large amount of cost data available on computer but they are not organized according to production line or product type. The numbers are gross purchases by month for particular materials and quarterly bills for services and utilities. The line operators who make up the dye baths do not follow any written specifications. They adjust the bath by adding more chemicals until the colors look right. This often results in a very rich mixture, which then flows down the drain after the job is complete. Interviews with the operators yield little else in the way of information, as even the addition of the chemicals may be done by sight or by scoopful, rather than by a measured quantity. Undaunted, we collect the following information and derive what is missing.

Cost Assumptions

1. Dye runs for the days that we are there to observe appear to consist of an average of seven chemicals. The names of the seven most common chemicals were determined and the gross annual purchase costs for each gathered from the purchasing department. These costs were divided among the number of process lines and weighted on basis of percent usage.
2. Treatment chemicals are used for pH adjustment (acid and caustic). Calculating the unit price of each, we use information from the maintenance man who monitors the pumps and meters regarding how often he changes the drums of the chemicals.
3. We use the company's labor rates for managers that perform safety training, and we fill out discharge monitoring reports to arrive at the regulatory compliance costs.
4. Under the current arrangement, there is no holdup of production when discharging to the sewer. The process tanks are drained and an equalization basin and mixer adjust the pH before the waters leave the facility.
5. The current system requires minimal maintenance. The pH meters are calibrated semiannually along with the flow meters.
6. Average utilities costs for dye processes are derived from utility bills and weighted according to the percentage of floor space the dye process occupies relative to other processes in the facility. Although this may overestimate or underestimate the actual amount of electricity and gas used, water use is fairly accurate because the dye process is the largest user of water.

Step 3: Defining Alternatives

Alternative 1: Installing a Precipitation Treatment System The first option considered by the team was the installation of an 800,000-gal/day flocculation, precipitation, and dewatering system. The company would then become a generator of filter cake, which is not expected to be hazardous provided metals loading can be held to a minimum. Under this option, additional chemicals would be introduced to the facility, utility usage would increase, and maintenance costs also would increase. There is a distinct possibility that production could be stopped if overflow alarms or pH balance alarms are triggered. Precautions would have to be taken to have the design approved by state and local officials.

We develop cost estimates for this process by talking to vendors of treatment systems and other manufacturers having similar-size daily flow rates.

Alternative 2: Eliminate Low-Quality Cloth The team feels that the largest contributors of metals come from the cheapest cloth suppliers. If the company were willing to spend a little more on material purchases, it may be possible to completely eliminate the problem without installing a treatment system. It also may be possible to negotiate with the suppliers and try to convince them to add less metal-based preservatives to the cloth.

Step 4: Conducting a Balanced TCA

Table 7.15 shows the relative costs of each project compared to the current system.

Cost Assumptions

1. Chemical purchases for production materials would not change under Alternative 1. The treatment chemicals would increase dramatically; see what follows. Under Alternative 2, the purchase cost of higher-quality cloth would increase, but this is allocated to production costs; see what follows.
2. Additional storage space will be required for both treatment chemicals and treatment wastes under Alternative 1. No increase is expected in Alternative 2.
3. Reporting and paperwork costs will increase as a result of Alternative 1.
4. Production costs and lost production time due to triggered alarm in the wastewater treatment plant could cost the company as much as $1,000/hr. According to other industries, these stoppages last about

TABLE 7.15 Financial Analysis of the P2 Project

Capital Costs	Description	Current Process	Waste Treatment	Better Fabric
Equipment purchase			$ 315,000	$ —
Disposal of old process			$ —	$ —
Research & design			$ 4,000	$ 2,000
Initial permits			$ 2,500	$ —
Building/process changes			$ 10,000	$ —
Total capital costs			$ 331,500	$ 2,000
Operating Costs	**Description**	**Current Process**	**Waste Treatment**	**Better Fabric**
Purchase costs	Chemicals	$ 23,000	$ 23,000	$ 23,000
Storage/floor space	Chemical storage	$ —	$ 400	$ —
	Waste storage	$ —	$ 1,500	$ —
Waste management	Treatment chemicals	$ 700	$ 2,700	$ 700
	Testing	$ 1,200	$ 1,500	$ 1,200
	Disposal	$ —	$ 4,500	$ —
Regulatory compliance	Safety/training equipment	$ 200	$ 250	$ 200
	Fees or taxes	$ —	$ —	$ —
	Manifests, test reports	$ 580	$ 1,000	$ 580
Insurance		N/A	N/A	N/A
Production costs	$/product	$ —	$ 2,000	$ 5,000
	Hours/product	$ —	$ —	$ —

(Continued)

TABLE 7.15 *(Continued)*

Maintenance	Time	$ 300	$ 13,000	$ 300
	Materials	$ —	$ 1,000	$ —
Utilities	Water	$ 560,000	$ 560,000	$ 560,000
	Electricity	$ 23,000	$ 25,000	$ 23,000
	Gas/steam	$ 12,000	$ 12,000	$ 12,000
Annual operating costs		620,980	647,850	625,980
Cash Flow Summary	**Description**	**Current Process**	**Waste Treatment**	**Better Fabric**
Total operating costs		$(620,980)	$(647,850)	$(625,980)
Incremental cash flow			$ (26,870)	$ (5,000)
−Depreciation			$ (33,150)	$ (200)
Taxable income			$ (60,020)	$ (5,200)
Income tax (40%)			$ —	$ 2,080
Net income			$ (60,020)	$ (7,280)
+Depreciation			$ 33,150	$ 200
After-tax cash flow			$ (26,870)	$ (7,080)
Present value	10 yr @ 10% = 6.1446		$(165,105)	$ (43,504)
Total capital cost			$(331,500)	$ (2,000)
Net present value			$(496,605)	$ (45,504)

10 min and occur roughly once per month. Alternative 2 will increase the costs of the raw cloth used in the dyeing process. This increase will run between $0.0025 and $0.0075/yd of fabric. The company processes an average of 1 million yd/yr of fabric.

5. Maintenance costs on Alternative 1 are anticipated to be $13,000 in labor and $1,500 in materials.
6. There will be some increase in utility costs in Alternative 1 (electricity for pumps, mixers, filter press) and none in Alternative 2.

Financial Analysis Conclusions

Neither option is a profit-making venture for the company. Both projects will hopefully prevent litigation by the sewer authority and also any penalties or fines. Alternative 2, working with suppliers and purchasing higher-quality cloth, appears to be the least costly alternative by a magnitude of 10:1. This option has the additional benefit of avoiding holdups during construction, additional regulatory burden, and potential for liability down the road with the filter cake.

Unique Features of This Case

It is important to note that the lack of hard data for either option introduces a great deal of uncertainty. In this case, the pollution prevention option (Alternative 2) saves the company from having to spend more to reach the same goal. It is possible that this option also will yield quality benefits resulting in fewer reject batches.

CASE STUDY 12: Circuit-Board Cleaning

The analysis calculates the maximum price per gallon for a profitable direct solvent substitute. This is accomplished by working numbers backward from a net present value slightly greater than zero. This technique could prove useful if you need to develop rough budget numbers for a project without having information about what the alternatives may be.

Step 1: Review of the Production Process

This small industry is a precious-metals job shop specializing in electrical component plating for corrosion resistance and conductivity. Low volumes of small parts are plated using hand racks. Parts come directly off the plating lines' final deionized water rinses to a water displacer and into a TCE degreaser. After cleaning, the parts dry in a drying closet and are packaged for shipping. The company desires to eliminate chlorinated solvents to save money and decrease labilities. The company does not know what alternatives are available and if they will work, but it would like to know the maximum budget that will allow it to break even. See Figure 7.10.

Step 2: Gathering Cost Data

The company does not have a computerized tracking system, although purchase records are stored in computer files. These files are not linked or searchable, requiring a time-consuming review of past records. Data gathering was further inhibited by a hands-off attitude of management and plant supervisors. They felt the consultant should be able to gather all the required information without any assistance from the staff. "That's what we're paying you for, right?" was the rhetorical comment.

Cost Assumptions

1. The current vapor degreaser is 4–5 years old. No salvage value was considered for this piece of equipment.
2. Chemicals used include a water displacer to remove water from the plated parts and TCE to remove the water displacer and provide spot-free drying. The use of both of these chemicals cost the company $22,000/yr.
3. Because the company uses TCE, it must file SARA 313 reports, state toxics-use-reduction reports, and pay a fee for the chemical.
4. Disposal costs are 5 drums of water displacer every 2 months (at $120/drum), and 2 drums of TCE waste every 2 months (at $150/ drum). Total disposal costs are $3,600/yr on the water displacer and $1,800/yr on the TCE. A direct substitution of TCE and the displacer most likely would result in some waste generated unless recycling equipment is considered. For this example, waste-disposal costs were kept the same to retain a conservative cost estimate.
5. Minimal maintenance is performed on the equipment. The water displacer tank is an old 55-gal drum with the top cut off. The TCE degreaser is cleaned out once a year.
6. No numbers were available for utility usage. The degreaser's name plate had been removed and the electrical ratings stamped on the

■ **FIGURE 7.10 Precious-Metal Plating Process Flow Diagram**

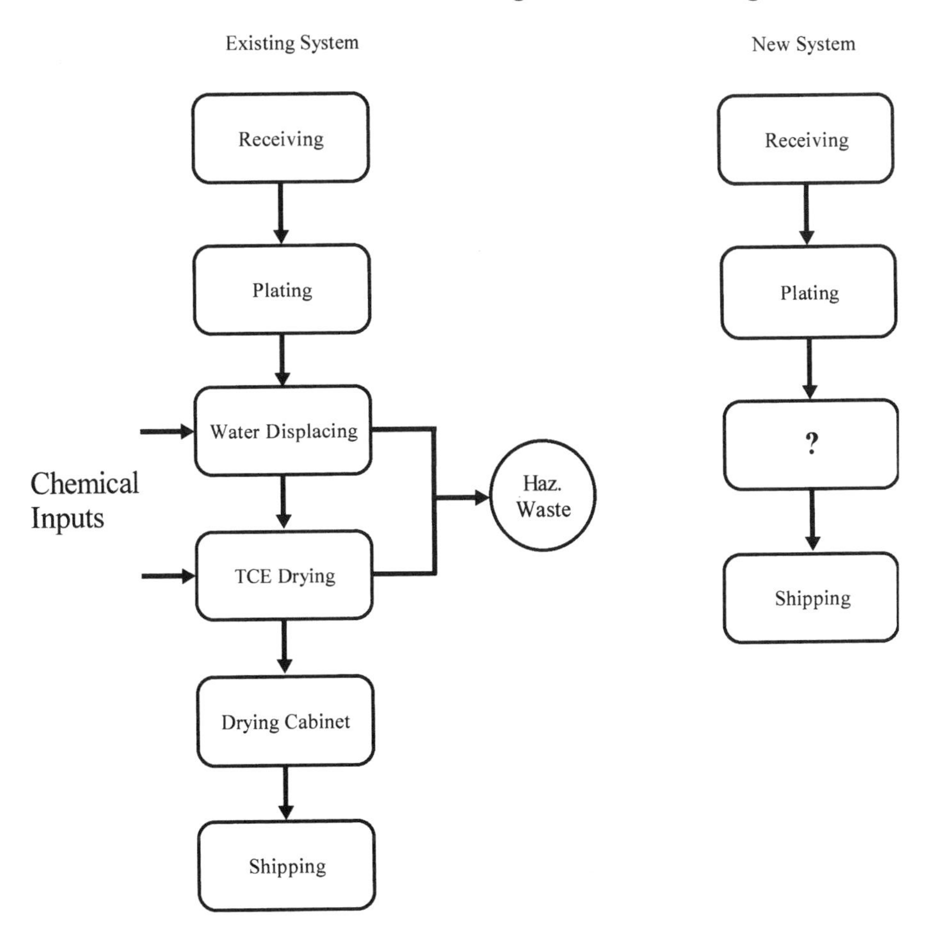

machine were not readable. Omitting this information will only make the budget estimate smaller, further ensuring a less costly alternative.

Step 3: Defining Alternatives

The goal is to develop a project budget for an alternative that will not exceed current operating costs. The amount of research needed was estimated because no meaningful research on alternatives had been done yet. Two alternatives illustrating both ends of the spectrum were chosen as bracketing examples and developed using industry average costs and data from other case studies.

Alternative 1: Drop-In Replacement The first option replaces TCE by using the same degreaser and water displacer. This option requires the least amount of change in process equipment and worker practices. Initial costs in this instance would be minimal and largely under the research-and-development category. Potential alternatives that could work as a drop-in would include terpenes, dilimonenes, alcohols, glycol ethers, and nonchlorinated hydrocarbon blends. Factors that narrow the options are flammability, acute and chronic toxicity, the need to have minimal residuals left on parts, and waste-disposal issues. The budget developed under these conditions used a 4-year lifetime for the project because regulations regarding parts-cleaning alternatives are constantly changing. New information regarding the toxicity of one or more of these alternatives may result in having to change to another cleaning solution.

Alternative 2: Elimination of the Water Displacer With additional process control, it may be feasible to completely eliminate the water displacer and the TCE degreaser by using hot, countercurrent, deionized water rinses, compressed air blowoff of parts, and drying in a forced-hot-air closet. Other electroplaters have achieved reliable spot-free drying using deionized water as the final rinse and then placing the parts in a hot-air drying booth. The company already has the deionized water system. This option requires the largest amount of change in process equipment and worker practices.

Step 4: Conducting a Balanced TCA

The TCA for the first alternative is based on a present value only slightly larger ($1) than the estimated capital cost for research. By working the present value calculations backward with a 4-year lifetime and 10% interest rate, we develop the after-tax cash flow, depreciation, and finally the total annual operating cash flow ($26,844). The following explains the assumptions made for this alternative.

Cost Assumptions

1. All costs were assumed to be the same as the existing system except for chemical purchase cost and fees or taxes. These assumptions are considered valid because this option requires the least amount of change. The water displacer would still be used and the vapor

degreaser would be used either as a vapor degreaser or as a heated cleaning tank. There is no reason to expect maintenance costs to change.

2. Fees or taxes would be eliminated provided the chemical chosen was not listed on EPA or state chemical reporting lists.
3. Chemical purchases were assumed to decrease by the incremental cash flow − fees or taxes. This decrease in purchase cost is necessary to compensate for the capital cost of research to find the alternative solvent and provide a break-even scenario.

Cost savings under Alternative 2 are significantly different. Research costs for this option were estimated at 1 wk of testing part cleanliness (40 hr × $20/hr). Initial equipment costs for this option were taken from a similar successful changeover from CFCs to hot-air drying. Modifications are made to the end of the plating line to reduce dragout from the final rinse and to blow off as much water as possible. These modifications cost $800 for piping, nozzles, and blowers. The drying closet is outfitted with hot air blowers at an additional cost of $10,000.[5] Process alterations consist of modifying the final rinse tanks on the plating lines that already have deionized (DI) water sources. The hot-air dryer was evaluated using a standard 7-year lifetime.

The second option completely eliminates the need for TCE and a water displacer, resulting in a large annual cost savings. Training costs are assumed to be the same and maintenance costs would be similar. The state chemical usage fee would be eliminated in this option as well.

Financial Analysis Conclusions

The third column in Table 7.16 lists the costs for the current system of drying parts using a water displacer and a TCE degreaser. The fourth column shows the not-to-exceed costs for an alternative solvent that would not cost more than the current system. The chemical must have a total annual purchase cost no greater than $15,950. If we knew for sure certain other characteristics of the alternative, such as its nonhazardous nature when disposing, then this cost could be increased by the amount saved from the waste-disposal cost category.

The last column in Table 7.16 shows the estimated costs for a project that completely eliminates purchases of TCE and a water displacer.

[5]Equipment costs based on similar successful solvent substitutions.

■ **TABLE 7.16 Financial Analysis of the P2 Project**

Capital Costs	Description	TCE & Displacer	Substitute & Displacer	Hot-Air Dryer
Equipment purchase			$ —	$ 10,000
Disposal of old process			$ —	$ —
Research & design			$ 3,200	$ 800
Initial permits			N/A	N/A
Building/process changes			$ —	$ 800
Total capital costs			$ 3,200	$ 11,600
Operating Costs	**Description**	**TCE & Displacer**	**Substitute & Displacer**	**Hot-Air Dryer**
Purchase costs	Chemicals			
	TCE or substitute	$16,000	$15,950	$ —
	Displacer	$ 6,000	$ 6,000	$ —
Storage/floor space	Chemical storage	$ —	$ —	$ —
	By-product storage	$ —	$ —	$ —
	Emissions storage	$ —	$ —	$ —
Waste management	Treatment chemicals	$ —	$ —	$ —
	Testing	$ —	$ —	$ —
	Disposal	$ 5,400	$ 5,400	$ —
Regulatory compliance	Safety/training equipment	$ 480	$ 480	$ 540
	Fees or Taxes	$ 1,100	$ —	$ —
	Manifests, test reports	$ 1,200	$ 1,200	$ —
Insurance		N/A	N/A	N/A
Production costs	$/Unit of product	N/A	N/A	N/A
	Labor/unit product	N/A	N/A	N/A
Maintenance	Time	$ 480	$ 480	$ 480
	Materials	$ —	$ —	$ —
Utilities	Water	N/A	N/A	N/A
	Electricity	Same	Same	Same
	Gas/steam	N/A	N/A	N/A
Annual operating costs		$30,660	$29,510	$ 1,020

(*Continued*)

■ **TABLE 7.16 *(Continued)***

Cash Flow Summary	Description	Current Process	Substitute & Displacer	Hot-Air Dryer
Total operating costs		$(30,660)	$(29,510)	$ (1,020)
Incremental cash flow			$ 1,150	$ 29,640
−Depreciation			$ (800)	$ (1,657)
Taxable income			$ 350	$ 27,983
Income tax (40%)			$ (140)	$(11,193)
Net income			$ 210	$ 16,790
+Depreciation			$ 800	$ 1,657
After-tax cash flow			$ 1,010	$ 18,447
Present value	4 yr and 7 yr		$ 3,202	$ 89,807
Total capital cost		$	(3,200) $	(11,600)
Net present value		$	2 $	78,207
Equivalent annual annuity	3.1699 and 4.8684		$ 1	$ 16,064

In this case, the capital cost for retrofitting the existing system could be as large as $110,000 and the project would still break even over the 7-year lifetime of the project. This maximum capital cost figure was back calculated by solving for the capital cost given the estimated savings and assumed lifetime.

The equivalent annual annuity had to be used to compare these two options due to their different lifetimes. By using the estimated capital costs shown in Table 7.16, the hot-air dryer would pay for itself in under 6 months and not be subject to future chemical restrictions.

Unique Features of This Case

This case study presents a unique method for understanding the relationship between capital costs and operating costs. In a budgeting exercise such as this one, each can be adjusted to maintain or affect the others.

CASE STUDY 13: Ion Exchange

A manufacturer has to clean tubes before they proceed to the assembly phases. Contaminants on the incoming tubes vary from road dirt and dust to condensation on the inside of the tubes from moisture buildup.

The pallets of tubes are not stored for extended periods of time and therefore, other contaminants from storage are minimal. Occasionally, the tubes arrive clean enough that they do not need washing.

Step 1: Review of the Production Process

The tubes are arranged vertically on racks and rolled into the washers. A single tank under the conveyor supplies heated deionized water that is pumped to the top of the washer and sprayed out an array of nozzles that corresponds in size and placement to the tubes. The washers run through an automatic cycle with fresh DI water being added to replace losses from spills, evaporation, and carryout on tubes and trays. The tubes then move down the line to the blower, which introduces hot air into the inside of the tube, alternating from top to bottom and bottom to top.

Overflows of rinse waters from the tube washer stations feed into a large centralized collection sump and then to the wastewater pretreatment system. Roughly, 80% of the 11 million gal/yr comes from washing over 50 million tubes. Because all this water must be deionized prior to use and then treated prior to discharge to the sanitary sewer, water conservation could be a great way to reduce costs for this process.

Hydrochloric acid (HCl) is used in the regeneration of ion-exchange columns in this operation. Air emissions of HCl are generated during the transfer of HCl from original containers to the wastewater treatment system and the deionized water system. The company estimates that 1% of the HCl used in these systems is emitted during transfer operations. Fugitive emissions were assumed to be 10% of the total emissions, with the balance going out the stack. Likewise, there are no data available on the efficiency of the exhaust system that removes fumes from the transfer area, but estimates are that 90% of HCl goes out the stack. Any water savings also would have a cascade effect on the hydrochloric acid, by decreasing the frequency of ion-exchange resin recharging and the use of this chemical. See Figure 7.11.

Step 2: Gathering Cost Data

Our interest in ion-exchange columns prompts us to investigate the costs associated with the tube washing process. Our first contact with staff in the facility was the process engineer, so we ask him questions as we tour the facility and during the debriefing session at the end of the

■ FIGURE 7.11 Process Flow Diagram for an Ion-Exchange System

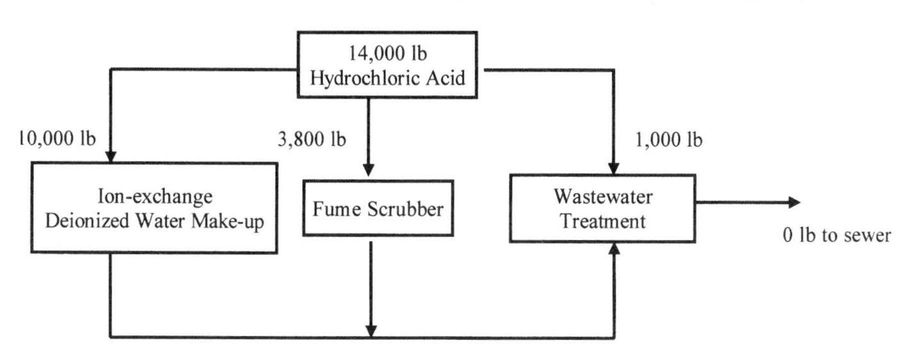

walk-through. Specifically, we need numbers on water and chemical use and costs and treatment and utilities requirements.

Because this process is not considered a part of any one particular product line, all the costs are allocated to overhead and maintenance. Few records exist other than chemical purchases and unit costs for sewer and water use. Rather than estimate these missing costs, we decide to leave them blank and see if the project will be feasible by just considering the larger ones.

Table 7.17 outlines the costs associated with the use of hydrochloric acid. Additional conditions used in determining the cost of the chemical are explained in what follows.

Cost Assumptions

1. Hydrochloric acid is stored in drums inside the facility. Storage and handling costs provided for other chemicals were applied to this situation. There are no storage costs for by-products or wastes associated with this chemical because it is neutralized in the wastewater treatment system.
2. There are no disposal costs associated with the use of hydrochloric acid because it is neutralized in the wastewater treatment process. Costs were unavailable for the quantities of sodium hydroxide used to neutralize the HCl.
3. Long-term liability connected with the use of hydrochloric acid is low. It is not considered a groundwater contaminant, carcinogen, mutagen, or teratogen. It is not an ozone precursor or depleting substance. Major risks during hydrochloric acid leaks are from fume

■ TABLE 7.17 Operating Cost for Using Hydrochloric Acid

Cost Item	Description	Cost Factor	Unit Price	Total Units	Total Cost
Purchase costs	Chemicals Hydrochloric acid	100%	\$ 0.85/lb	14,203 lb	\$12,072.00
Storage floor space	Ancillary storage	—	—	—	\$ 0.00
	Chemical storage	100%	\$15.50 ft^2	24 ft^2	\$ 372.00
	By-product (scrap) storage	0	0	0	\$ 0.00
	Emissions (waste) storage	0	0	0	\$ 0.00
Waste management	Treatment chemicals	0	0	0	\$ 0.00
	Testing	—	—	—	N/A
	Disposal	0	0	0	\$ 0.00
Regulatory compliance	Safety/training equipment	—	—	—	N/A
	Fees & taxes				
	Base fee	14%	\$4,625.00	—	\$ 647.50
	Chemical	100%	\$1,100.00	—	\$ 1,100.00
	Manifests, test reports	—	—	—	N/A
Insurance		—	—	—	N/A
Production costs	\$/Unit of product	0	0	0	\$ 0.00
	Labor/unit product	0	0	0	\$ 0.00
	Other	0	0	0	\$ 0.00
Maintenance	Time	—	—	—	N/A
	Materials	—	—	—	N/A
Utilities	Water	—	—	—	N/A
	Electricity	—	—	—	N/A
	Gas/steam	—	—	—	N/A
Total annual operating cost					\$14,191.00

exposure to mucous membranes and burns from dermal contact. Because of this limited risk potential, insurance costs are not a factor in this analysis, and the impact of public perception is also a low-weighted factor.

4. Production and maintenance costs were unavailable for this chemical. No costs are assigned to the use of hydrochloric acid in regenerating ion-exchange beds. Wastewater treatment costs do not apply in this calculation.

Step 3: Defining Alternatives

All of the tube washing water must be deionized to remove metals, minerals, and salts present in the city water. Iron and copper are the two largest contaminants removed by ion exchange. By reducing the amount of water purchased from the city, the company will save money, conserve resources, and use the ion-exchange system less. This provides a direct reduction in use of hydrochloric acid and sodium hydroxide to regenerate ion resins, thereby reducing chemical purchase costs and potentials for worker accidents involving hydrochloric acid.

Alternative 1: Filtration and Reuse The current arrangement of a centralized sump provides a collection point where these used wash waters could be returned to the ion-exchange columns for additional cleansing. Additional filters may be required to separate fines and soil to allow complete reuse. The purchase and maintenance on these filters would be minimal compared to the current operating costs for the once-through wash system. The company potentially could reduce purchases of city water (for this operation) by 90% using this system.

Alternative 2: High-Pressure Nozzles In addition to reusing wash-waters, the company could further reduce water demand by replacing the existing nozzles with high-pressure spray nozzles. Water reductions of up to 80% have been achieved through the use of these nozzles. The investment in 392 new nozzles could reduce water flow by 30% with no decrease in the quality of rinsing. Based on current water use, the company could save 3.1 million gal/yr by converting to water-conserving nozzles.

Step 4: Conducting a Balanced TCA

We gathered additional data on costs for Alternative 2 by consulting with vendors of high-pressure nozzles and contractors that could construct a water return system. The company may choose to refit the system with its own workers, in which case the labor costs may come down.

The following and Table 7.18 detail the financial feasibility of reusing the rinsewaters that are collected in the sump and retrofitting the washing stations with new nozzles. In this case, rinsewaters are collected and piped back to the DI columns for filtering and reuse. These two projects represent a 33% decrease in water use running through the system, and a 90% decrease in purchases of water from the city.

Cost Assumptions

1. The numbers in Table 7.18 do not consider the potential savings and reduction in use of sodium hydroxide. Additional savings would result from decreased freshwater deionization, including extending the life of the ion-exchange resins, maintenance costs, and electricity to run the pumps and heaters.
2. The disposal cost in this instance represents a sewer usage fee attached to the discharge of wash waters. The company is charged \$2.23/100 ft^3 of water discharged. An estimated 9.4 million gal of effluent is connected to the use of the water deionization and tube washing system. This costs the company \$28,145/yr.
3. The utilities cost for water is the cost of the 9.4 million gal estimated to be deionized and used in the tube washing operation. At \$2.06/100 ft^3, water usage costs the company \$26,000/yr.
4. The numbers presented here represent estimates on time and materials required to repipe the rinse waters from the collection sump into the deionization columns and retrofitting the washing machines with new nozzles. There are nine washing machines in the facility. Once a workable technique is developed, applying it to the other machines becomes easier. Equipment to be purchased would be additional pipe and fittings, perhaps a recirculating pump. Building and process change costs would consist of labor to modify the piping and to connect a new pump if necessary. Research costs consist of time to find the appropriate pumps and to perform a quality check to ensure no decline in washing cleanliness.
5. HCl is anticipated to decrease by 33% in direct relation to the amount of water that requires deionizing. If one-third less water is being drawn from the city, then one-third less water requires deionization. HCl costs \$0.85/lb and equates to a savings of \$4,024/yr.

■ **TABLE 7.18 Financial Analysis of the P2 Project**

Capital Costs	Description	Current Process	Rinse Modification
Equipment purchase			5,000
Disposal of old process			
Research & design			2,000
Initial permits			
Building/process changes			2,000
Total capital costs			9,000
Operating Costs	**Description**	**Current Process**	**Rinse Modification**
Purchase costs	Chemicals HCl	12,072	8,048
	Ancillary	0	0
Storage/floor space	Chemical storage	372	372
	By-product storage	0	0
	Emissions storage	0	0
Waste management	Treatment chemicals	0	0
	Testing	N/A	N/A
	Disposal	28,145	18,857
Regulatory compliance	Safety/training equipment		
	Fees & Taxes		
	Base fee	648	648
	Chemical fee	1,100	1,100
	Manifests, test reports	N/A	N/A

(Continued)

■ TABLE 7.18 *(Continued)*

Operating Costs	Description	Current Process	Rinse Modification
Insurance (2)		N/A	N/A
Production costs	$/Unit of product	0	0
	Labor/unit product	0	0
	Other	0	0
Maintenance	Time	N/A	N/A
	Materials	N/A	N/A
Utilities	Water	26,000	17,420
	Electricity	N/A	N/A
	Gas/steam	N/A	N/A
Annual operating costs		68,337	46,445
Cash Flow Summary	**Description**	**Current Process**	**Rinse Modification**
Total operating costs		(68,337)	(46,445)
Incremental cash flow			21,892
−Depreciation			(900)
Taxable income			20,992
Income tax (40%)			(8,396)
Net income			12,595
+Depreciation			900
After-tax cash flow			13,495
Present value	10 yr @ 10% = 6.1446		82,922
Total capital cost			9,000
Net present value			73,922

6. Disposal costs (the sewer usage fee) would decrease by one-third, saving the company $9,287/yr.
7. A one-third decrease in HCl usage may also decrease the company's regulatory burden, by dropping this chemical's usage below the reporting threshold for the state pollution prevention planning and reporting law. Implementing this water reuse project with other pollution prevention projects would ensure that HCl usage stays well below chemical reporting thresholds. Despite this possible savings, this analysis retained the chemical use fee to represent a conservative estimate of savings.
8. Utility costs decrease to one-third less water usage, saving over $8,580/yr.

Financial Analysis Conclusions

This project results in a net savings of $73,000 over a 10-year lifetime. Several costs were not discounted in this analysis to portray a more conservative project. Even without these additional savings, such as reduced sodium hydroxide usage, the project looks financially feasible.

CASE STUDY 14: Slot Coating

An international manufacturer of camera and X-ray film uses large slot-coating machines to deliver solutions to a "web." Various chemicals are used as hydrolyzing agents and carriers for coating formulations. These formulations are applied to the web, or film material, and dried to create either a finished or intermediate product. The wastes from the process include water solutions and fumes, both of which are treated on-site. Figure 7.12 shows the major process steps involved in the coating operations and emissions created.

Step 1: Review of the Production Process

This process uses sodium hydroxide as a component mixture to the coatings and as pH adjustment in the fume scrubber. Virgin chemical is blended with methanol in the mixing drum. The mixed solution is pumped to the hydrolizer and applied to the web. Average throughput rates for web and coating substrates exceed 6,000,000 ft^2/yr, resulting in a low ratio of sodium hydroxide usage per square foot of coated web.

■ **FIGURE 7.12 Process Flow Diagram for Slot-Coating Process.**

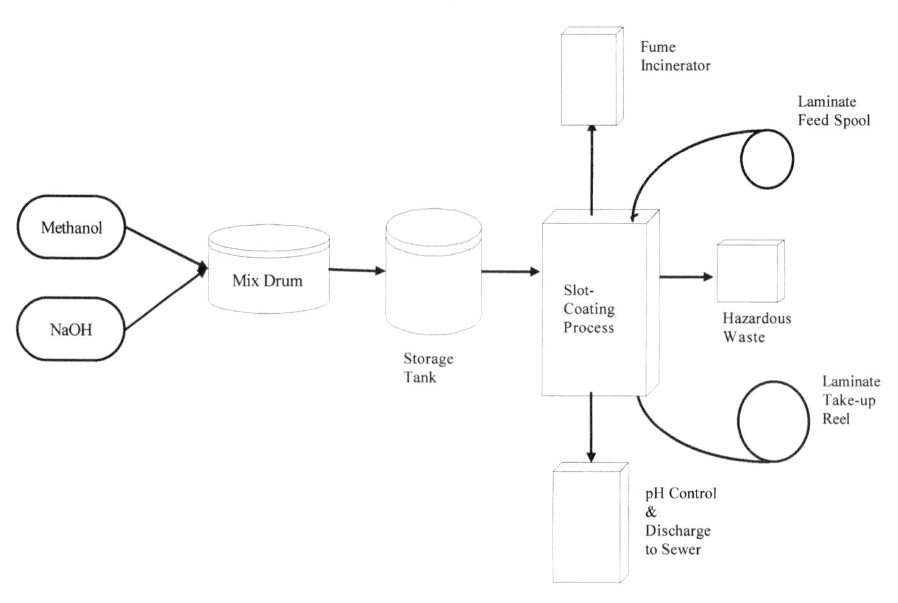

By-products from this process include roughly 2000 lb/yr NaOH within the wastewaters sent to the pretreatment system and another 5000 lb/yr leaves the production unit as hazardous waste. Due to the quantities of this chemical used by the slot-coating process, the company's pollution prevention team desires to reduce usage to save money, prevent pollution, and decrease the regulatory burden associated with this process.

Step 2: Gathering Cost Data

Fortunately for us, the process engineer had all the data required available on a computerized materials tracking database. He was able to pull up figures quickly from past years and current year purchases to show trends in chemical use and cost increases. Costs associated with the use, processing, and disposal of sodium hydroxide are detailed in Table 7.19. The assumptions and conditions effecting cost data are described in what follows.

1. Purchase costs include the charge for the drums needed to ship and receive sodium hydroxide. This reflects the completeness of the computerized materials tracking system.
2. Costs were unavailable for electricity usage because laminating machines are not metered separately from facility lighting and other

TABLE 7.19 Operating Cost for Using Sodium Hydroxide

Cost Item	Description	Cost Factor	Unit Price	Total Units	Total Cost
Purchase costs	Chemicals 50% NaOH	100%	$ 0.11/lb	13,624 lb	$1,498.60
	Ancillary Drums	35%	$ 9.70/drum	36 drums	$ 122.22
Storage/floor space	Chemical storage	100%	$ 25.85/ft^2	16ft^2	$ 413.60
	By-product (scrap) storage				
	Emissions (waste) storage	35%	$ 25.85/ft^2	8ft^2	$ 72.38
Waste management	Treatment chemicals	100%	$ 0.97/lb	330.6 lb	$ 320.80
	Testing	35%	$ 100/sample	5	$ 175
	Disposal	35%	$ 210/drum	36	$ 2,646
Regulatory compliance	Safety/training equipment	35%	$ 100/person	10	$ 350
	Fees (i.e., TURA)	100%	—	—	$ 1,100
	Manifests, test reports	10%	$ 11,430	—	$ 1,143
Insurance		N/A	N/A	N/A	N/A
Production costs	$/Unit of product	—	$ 0.03 linear ft	2.44×10^6 linear ft	$ 73,231
	Labor/unit product	—	$ 0.03 linear ft	2.44×10^6 linear ft	$ 73,231
	Other	N/A	N/A	N/A	N/A
Maintenance	Time	35%	$ 40/hr	80 hr	$ 1,120
	Materials	35%	$ 2,000		$ 700
Utilities	Water	35%	$1,000.001885/gal	624,000 gal	$ 411.70
	Electricity	N/A	N/A	N/A	N/A
	Gas/steam	N/A	N/A	N/A	N/A
Total annual cost of toxics					$ 156,535

motors. We did not develop estimates due to the short timeline and limited budget of the project.

3. Contingent liability connected with the use of sodium hydroxide is low. It is not considered a groundwater contaminant, carcinogen, mutagen, or teratogen. It is not an ozone precursor or ozone-depleting substance. As a reactive chemical with corrosive properties, the largest risks are to the workers handling the sodium hydroxide. Data on accidents involving sodium hydroxide were not collected as part of this analysis because of time and budget limitations. Insurance premiums do not reflect the risk of using this chemical and would not change if its use were eliminated. Because of the low risk potential, insurance costs are not a factor in this analysis.

Step 3: Defining Alternatives

Options for reducing or eliminating the use of this chemical in slot-coating processes were discussed at a regular pollution prevention team meeting. The following options were considered worthy of financial analysis and pursuit if profitable.

Alternative 1: Changing Product Structure The team discussed altering the chemical makeup of the products' substrate, which in turn would change the need for hydrolysis and decrease the usage of sodium hydroxide. The reformulation is technically feasible through careful alterations in process chemistry and research. These efforts would take at least 2 years, but could reduce sodium hydroxide use by 10%.

Alternative 2: Changing the Lamination Process The second process change involves modernizing the slot-coating machines and substrate to allow hydrolyzing on one side only. Research currently underway confirms the technical feasibility of this option. The company estimates full implementation of this project within 5 years and estimates sodium hydroxide use will be reduced an additional 10%. A financial analysis of this option follows.

Step 4: Conducting a Balanced TCA

The pollution prevention team agreed that both changes would be implemented together even if only one of them were profitable. Therefore, the reductions shown in the analysis reflect the total cost reductions by implementing both alternatives. The assumed reduction in sodium hydroxide use is 20% with a 10-year lifetime for all innovations and process changes. Table 7.20 details the TCA calculations and incremental cash flows.

Cost Assumptions

1. This analysis estimates a 20% reduction in NaOH purchases and corresponding reductions in the number of drums and storage space required. Similar reductions in wastewater treatment cost and hazardous-waste disposal are included.
2. Although there may be some production cost savings from reducing the need to hydrolyze the substrate, this cost was held static in case the new formulation also resulted in a longer drying time or slower application speeds.
3. Research-and-development costs are not allocated to specific production centers, but handled as a separate cost category and borne by the corporation as a whole. Because of this, it was difficult to estimate the amount of research conducted to date and needed before implementation of the projects. A rough estimate of $42,000 was included to provide some leeway for the calculation even though no actual chargeback will be made to the process.

Financial Analysis Conclusions

Research and development was the largest capital cost and it could be debated whether research is an actual capital cost. If taken as such, the project is not financially viable. Over the 10-year life span, the changes would cost over $38,000. It is probably more accurate to assume that the research costs will not be considered a depreciable expense and therefore only the process changes would be considered. If this is the case, the project has a net present value of over $4,000. Although this is a positive value, a higher value would create more enthusiasm for implementation. Contingent costs were not included here, but given that over 5000 lb/yr of sodium hydroxide leaves the facility as hazardous waste, these costs could be considerable.

CASE STUDY 15: Electronic Circuits

This medium-size manufacturer produces military and civilian electronic equipment such as personal identification coders, infrared surveillance

■ TABLE 7.20 Financial Analysis of the P2 Project (1993)

Capital Costs	Description	Current Process	Less Hydrolysis
Equipment purchase			$ 0
Disposal of old process			$ 0
Research & design			$40,000
Initial permits			$ 0
Building/process changes			$ 2,000
Total capital costs			$42,000
Operating Costs	**Description**	**Current Process**	**Less Hydrolysis**
Purchase costs	Chemicals		
	NaOH	$ 1,499	$ 1,198
	Ancillary Drums	$ 122	$ 98
Storage/floor space	Chemical storage	$ 414	$ 331
	Byproduct storage	$ 0	$ 0
	Emissions storage	72	58
Waste management	Treatment chemicals	$ 321	$ 257
	Testing	$ 175	$ 175
	Disposal	2,646	$ 2,117
Regulatory compliance	Safety/training equipment	350	350
	Fees or Taxes		
	Base chemical fee	$ 1,100	$ 1,100
	Chemical fee		
	Manifests, test reports	$ 1,143	$ 914
Insurance		N/A	N/A
Production costs	$/Unit of product	$73,231	$73,231
	Labor/unit product	$73,231	$73,231
Maintenance	Time	$ 1,120	$ 894
	Materials	$ 700	$ 560

(Continued)

■ **TABLE 7.20** *(Continued)*

Operating Costs	Description	Current Process	Less Hydrolysis
Utilities	Water	$ 412	$ 412
	Electricity	N/A	N/A
	Gas/steam	N/A	N/A
Annual operating costs		$150,719	$154,868
Cash Flow Summary	**Description**	**Current Process**	**Less Hydrolysis**
Total operating costs		(150,719)	(154,868)
Incremental cash flow			(4,149)
−Depreciation			(4,200)
Taxable Income			(8,349)
Income tax (40%)			(3339.6)
Net income			(11,689)
+Depreciation			(4,200)
After tax cash flow			(7,489)
Present value	10 yr @ 10% = 6.1446		(46,014)
Total capital cost			(42,000)
Net present value			(88,014)

systems, location and position monitors, and global positioning sensors. Due to the high voltages and low currents of these electronic devices, soil contamination must be kept at a minimum or result in product failure.

The company began looking for alternatives to Freon and purchased an ultrasonic cleaning system and began searching for replacement solvents. In the interim, the corporate directive came down to eliminate usage of CFC 113. Hydrochloroflurocarbon (HCFC) 141b was quickly brought in as a drop-in replacement for Freon 113. The company manufactured customized bench-top vapor degreasers just for HCFC 141b.

This case takes a retrospective look at the costs for changing from CFC 113 to HCFC 141b, and then the potential costs for changing from HCFC 141b to a terpene alternative.

Step 1: Review of the Production Process

The circuit boards are assembled at benches by workers using soldering irons and small hand tools. These board assemblies have solder and rosin on them that must be cleaned off. Prior to eliminating CFCs, the workers had to bring their assemblies to a centralized vapor degreaser and to allow the degreaser operator to clean their boards. With the current HCFCs, each two assembly people have a small degreaser. A maintenance person comes by and refills each bench-top degreaser and makes sure it is in working order. See Figure 7.13.

Step 2: Gathering Cost Data

The company's entire inventory and materials requisitioning process is computerized. This system also generates reports on annual usage by product or by material type. Extracting information for the TCA was relatively simple, requiring a few minutes of one of the design engineer's time. These data and conversation with the maintenance staff and design engineers provided as much data as were available for the two TCAs.

1. Year-to-year usage of Freon varied due to contracts and the recession. The company used an average of 14,000 lb of Freon 113 per year. At the time the project was conducted, the market price for CFCs was $8.57/lb, up from $5.00/lb in previous years.
2. The company purchases 1000 lb/mo of HCFC 141b (12,000 lb/yr × $3.15/lb = $37,800/yr).
3. Disposal costs for the CFC and HCFC run the same at $150/yr.
4. The company estimates that filing paperwork related to the use of CFCs is 16 hr/yr × $25/hr = $400/yr. The company 's only reportable chemical was CFC 113. Because of the change to HCFC 141b, the company will not have to develop a state waste minimization plan. These plans typically require an average of 4 weeks of intensive work by a team of several employees. Not having to develop a plan results in approximately $8,000 of savings (2 persons × 40 hr/wk × 4 wk × $25/hr). This savings was distributed over the lifetime of the alternatives (2 years for HCFC and 7 years for the alternatives yet to be discussed) and appears as a negative cost in the regulatory compliance category.

FIGURE 7.13 Printed Circuit-Board Cleaning System

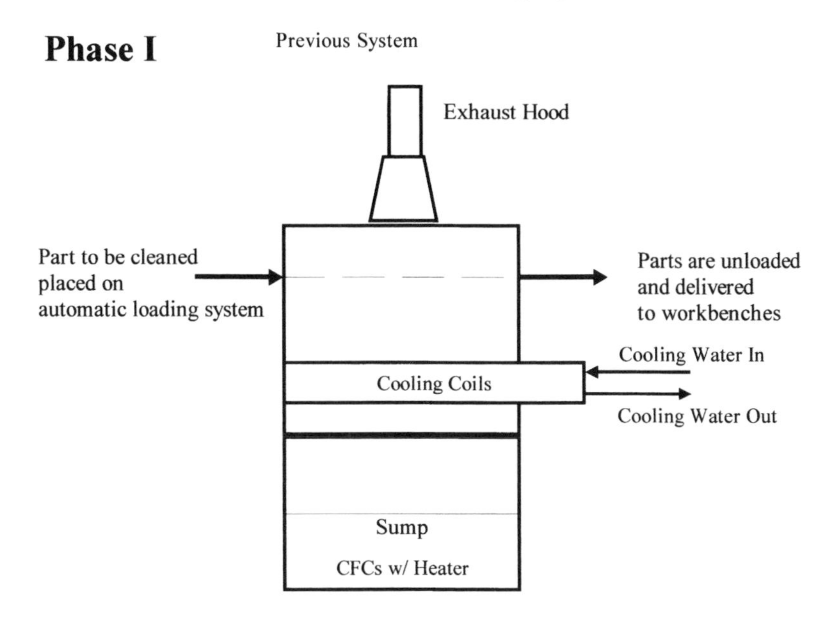

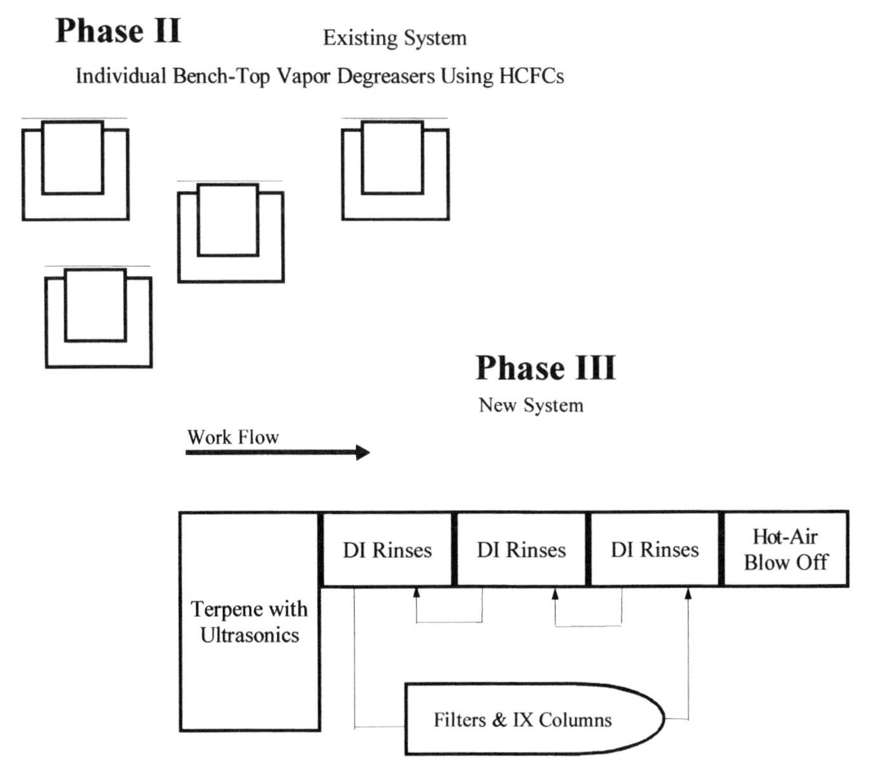

5. Worker health and safety training attributable to CFCs costs $50/yr. Safety training for the HCFC consisted of writing a new cleaning procedure and circulating a memo.
6. Maintenance on the old CFC degreaser was 4 hours every 6 months (8 hr/yr). Electrical usage was estimated using a standard vapor-degreaser specification on file: 230 V, 60 A.

Step 3: Defining Alternatives

The company was concerned with maintaining the reliability of its product and hence the level of cleanliness achieved with CFCs. The switch to HCFCs was quick and required little testing to maintain quality standards.

Alternative 1: Bench-Top Degreasers The custom-built bench-top degreasers took two engineers six months to design and develop the solid-state controls (6 mo × 4 wk × 40 hr × 2 people × $25/hr = $48,000). Manufacturing the units required 60 hr/unit × $8.50/hr × 18 units = $9,180. The existing solvent distillation system was also modified to recycle HCFCs ($453). No permits or building changes were required for the switch to HCFCs. The company will use HCFCs for 2 years while they find a suitable chemistry for their ultrasonic unit.

Alternative 2: An Ultrasonic System The ultrasonic system consists of a single immersion tank with ultrasonics followed by heated, three-stage, countercurrent rinse tanks with ultrasonics in the final rinse and a forced-hot-air dryer. Initial cost for the unit was $25,525. The new ultrasonic unit required raising the suspended ceiling and new water and discharge lines ($1,009). Estimates on research required are 1/4 of a full-time employee for 1 year ($15/hr × 500 hr/yr = $7,500/yr). The unit is given a 7-year lifetime.

Labor rates are for maintenance ($15) and paperwork ($25). The cost of capital is 10%.

Step 4: Conducting a Balanced TCA

1. The current bench-top systems allow workers to clean their own parts without waiting and do not require an extra person to operate a vapor degreaser. This results in a savings in production costs.

2. Initially, the individual HCFC degreasers required replacement valves every 2 months. The engineering team resolved this issue and the replacements are considered a one-time maintenance cost and have been added to the equipment purchase category under this option (two valve replacements per machine every 2 months × 15 machines × $6/valve and taking 15 min/valve).
3. The bench top units come in three sizes, each with a different electrical demand: (twelve) 1.1-gal units at 300 watts each; (two) 2.55-gal units at 500 watts each; and (four) 5-gal units at 700 watts each. The cost of electricity is $0.05/kWh. Total electrical demand for these machines running 8 hr/day, 250 day/yr is $740. Each unit is drained at the end of the day, so the machines need not run continuously to limit evaporative losses. This is estimated to take 2 min/day/machine and costs the company $1,875/yr.
4. Current market prices for terpenes range from $12–$15/gal. If the 8-gal UFI sump under heavy use required changing every month, it would mean 96 gal of use and waste, or $1,400/yr in chemical costs and approximately $330 in disposal costs.
5. Filters for the ultrasonic system need to be replaced in the following frequency: the $44 and $100 types every month and the $5 type every week. The ultrasonic unit's electrical demand is 240 V at 36 A, or $1,209/yr. Water use is nominal due to the deionized rinses that continually recirculate from the tanks to the DI columns and back. The columns have not been recharged yet, so no cost data were available.[6]

Table 7.21 details the TCA calculations and incremental cash flows.

Financial Analysis Conclusions

The calculations show $163,228 annual operating costs for the CFC vapor degreaser. Switching to bench-top style units, manufactured in-house and equipped with HCFCs, saved the company $87,408/yr. Although this may appear to be a significant cost savings, 34% of that savings comes from eliminating the one worker who operated the vapor degreaser. The company claims productivity may have increased because assembly workers can clean parts at their convenience without

[6]The author gratefully acknowledges the Massachusetts Toxics Use Reduction Institute, Lowell, Massachusetts, for assistance with Case Studies 7, 12, 15 and 16.

■ **TABLE 7.21 Financial Analysis of P2 Project**

Capital Costs	Description	Old CFC Process	Current HCFC	New Ultrasonic
Equipment purchase			$ 9,810	$ 25,525
Disposal of old process			In storage	In storage
Research & design			$ 48,000	$ 7,500
Initial permits			$ —	$ —
Building/process changes			$ 440	$ 1,009
Total capital costs		$ 58,250	$ 34,034	
Operating Costs	**Description**	**Old CFC Process**	**Current HCFC**	**New Ultrasonic**
Purchase costs	Chemicals			
	CFCs	$120,348	$ —	$ —
	HCFCs	$ —	$ 37,800	$ —
	Terpenes	$ —	$ —	$ 1,400
Waste Management	Treatment chemicals	$ —	$ —	$ —
	Testing	$ —	$ —	$ —
	Disposal	$ 150	$ 150	$ 330
Regulatory compliance	Safety/training equipment	$ 50	$ 50	$ 50
	Fees or taxes	$ 9,700	$ —	$ —
	Manifests, test reports	$ 400	$ (3,600)	$ 742
Insurance		N/A	N/A	N/A
Production costs	$/Unit of product	$ —	$ —	$ —
	Labor/unit product	$ 30,000	$ —	$ 30,000
Maintenance	Time	$ 120	$ 1,875	$ 108
	Materials	$ —	$ —	$ 1,978
Utilities	Water	$ —	$ —	Nominal
	Electricity	$ 2,460	$ 740	$ 1,209
	Gas/steam	$ —	$ —	$ —
Annual operating costs		163,228	$ 37,015	$ 35,817

(*Continued*)

■ **TABLE 7.21** *(Continued)*

Capital Costs	Description	Old CFC Process	Current HCFC	New Ultrasonic
Total operating costs		$(163,228)	$ (37,015)	$(35,817)
Incremental cash flow			$ 126,213	$ 127,411
−Depreciation			$ (29,125)	$ (4,862)
Taxable income			$ 97,088	$ 122,549
Income tax (40%)			$ (38,835)	$(49,020)
Net income			$ 58,253	$ 73,529
+Depreciation			$ 29,125	$ 4,862
After-tax cash flow			$ 87,378	$ 78,391
Present value	2 and 7 yr		$ 151,644	$ 381,641
Total capital cost			$ (58,250)	$(34,034)
Net present value			$ 93,394	$ 347,607
Equivalent annual annuity			$ 53,814	$ 71,401

waiting. On the other side, chemical use per part is probably higher because of the difficulty of having all workers to raise and lower their parts at a correct rate, empty and refill their bench-top units carefully, and keep them covered when not in use.

The ultrasonic system provides a larger net present value, but the two projects cannot be compared side by side until adjusting for the effect of their different lifetimes. In this case, the ultrasonic system provides a higher annual return over its lifetime than the HCFC system.

Unique Features of This Case

It is important to note the difficulty of comparing events that occurred in the past with those that occur now or will occur in the future. Current prices for all cost categories must be obtained. Sometimes there are no data on past or current prices. Other times, the prices have changed so dramatically that it may seem unreal. The change in price of CFCs is a prime example. The only way to obtain CFCs within the United States is to buy recycled CFCs at a premium. Just 10 years ago they were one of the least expensive solvents, widely used in everything from air conditioners to foam insulation and industrial blowing agents.

Also note that the profitability of Alternative 2 was maintained even with rehiring the degreaser operator. This may be another untapped benefit of pollution prevention.

CASE STUDY 16: Disposal Options

This case study explores the costs associated with three common disposal alternatives that face companies using aqueous and semiaqueous cleaning chemistries. Oftimes, the costs for disposing or treating an additional waste stream are not considered and can have a significant impact on the profitability of the project.

Step 1: Review of the Production Process

Operations at this 15-employee manufacturer of precision metal parts consist of milling, tapping, reaming, and close-tolerance lathe work. Most of the parts are small precision components (gears, pinions, etc.) and ordered in large quantity. Parts are machined brass, aluminum, stainless, or cast metals and often have threaded holes and recesses. Soils are either petroleum-based threading and cutting lubricants or water-based machine-tool coolants. The company's other wastes consist of scrap metal and spent water-based cutting fluids. The company president ordered a change from 1,1,1-trichloroethane to an aqueous cleaning solution, but now it must decide how to best dispose of the spent cleaning solution. A TCA will be performed to assist in making a decision. See Figure 7.14.

Step 2: Gathering Cost Data

The company president keeps paper records of all purchases and also has substantial recall ability as he is the prime decision maker. Subsequently, most data needed for the analysis were readily available.

Last year, the old vapor degreaser used 2500 lb of solvent for a total purchase cost of $2,400. Off-site disposal of 668 lb cost $150 that year. The company purchased an aqueous cleaning system consisting of two separate tanks, one wash and one rinse, for $10,000. The old vapor degreaser is still on-site, in storage. Chemical purchase costs for the new system are about $75/yr. The unit has only used half of a 55-gal drum in the first year of operation. Each drum costs $150.

FIGURE 7.14 Precision Parts Cleaning Process

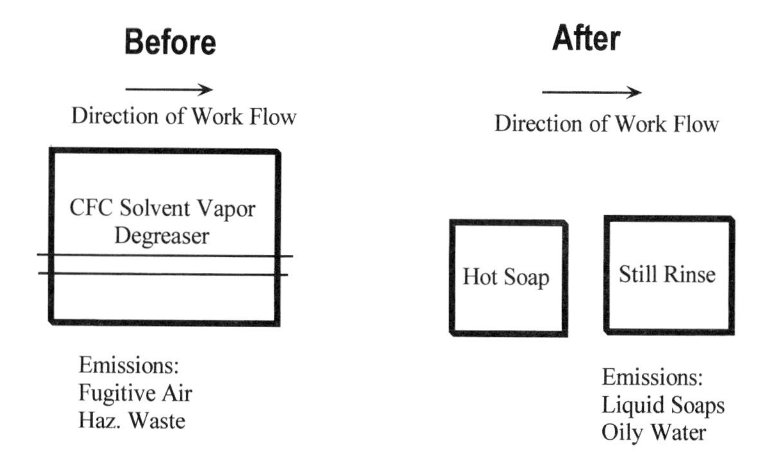

1. This facility does not use enough chemicals to break the thresholds for filing under federal release reporting laws.
2. The company found processing time takes longer with the new system. It effectively cleans only 80% of the parts. The remaining 20% require hand cleaning, mostly due to blind holes. The extra time required is not billable, that is, the parts can be left in the washer and the worker can move on to some other task.
3. The in-house labor rate is around $20/hr given health insurance and benefits. However, if workers could be billed under a contract, they would bring in much more than that on contract.

Step 3: Defining Alternatives

Alternative 1: Sending the Waste Off-Site The first option is to drum all the waste cleaner and have a contractor come and remove the drum on a semiannual basis. The quantities are so small that the company will not change the waste generator status.

Alternative 2: Discharging the Waste to the Sewer Another possibility for the company is to discharge the spent cleaning solutions, which are mostly water, to the sewer. The company currently does not have a water discharge except for sanitary water. This would require obtaining a general permit for oily wastewaters and setting up a small treatment system to remove oils and adjust pH.

Alternative 3: Recycle Using an Ultrafiltration System The final option is to purchase a small ultrafiltration system and separate the water from the oils and soaps. The water could be recycled and a much smaller portion of waste would be generated. The company also would be able to recycle its water-based machine coolants, which are currently sent off-site for disposal.

Step 4: Conducting a Balanced TCA

Each of the three options is detailed in Table 7.22. The assumptions for these costs follow.

Sending the Waste Cleaner Off-Site An annual laboratory testing fee of $400 was included to assure the waste hauler of the exact specifications of the wastes leaving the facility. Disposal costs were determined as follows: The facility has roughly 150 gal of coolant in machine sumps and replaces it once per year. Disposal of waste coolants average $100/drum, or $300 per year. Alkaline cleaning solutions average $35/drum with the cleaning tank emptied six times per year. Paperwork connected with testing and manifesting the waste was figured at 1 hour every 3 months at $20/hour. It is important to note the facility would change hazardous-waste status under this option from a very small quantity generator to a Small quantity generator

Discharging the Wastes to the Sewer A $300 laboratory test fee was included for each batch dump to the sewer. Although not specifically required by law, the facility is located next door to the municipal water-treatment plant and additional testing was deemed a show of good faith and possible protection against future legal actions. Acid would have to be maintained on-site to balance the pH of the wastes before discharge. A half hour of paperwork is connected with this option for each test and discharge.

Recycling the Wastes The ultrafilter would cost $3,000 and has an expected lifetime of only 8 years. There is additional maintenance connected with the ultrafilter and the concentrate still must be manifested off-site. Sludge from the ultrafilter would be sent off-site at a cost of $70/yr. Having an ultrafilter on-site for recycling the cleaner would also allow the company to run water-soluble machine coolants through the

TABLE 7.22 Financial Analysis of the P2 Project

Capital Costs	Description	Old Process	Ship Waste Off-Site	Discharge to Sewer	Recycle w/ UF
Equipment purchase			$ —	$ —	$3,500
Disposal of old process			In storage	In storage	In storage
Research & design			$ —	$ —	$ —
Initial Permits			$ —	$ 600	$ —
Building/process changes			$ 300	$ 400	$ —
Total capital costs			$ 300	$1,000	$3,500
Operating Costs	**Description**	**Old Process**	**Ship Waste Off-Site**	**Discharge to Sewer**	**Recycle w / UF**
Purchase Costs	Chemicals				
	TCE	$2,700	$ —	$ —	$ —
	Alkaline detergent	$ —	$ 75	$ 75	$ 50
Waste management	Treatment chemicals	$ —	$ —	$ 38	$ —
	Testing	$ —	$ 400	$1,800	$ —
	Disposal	$ 167	$ 545	$ —	$ —
Regulatory compliance	Safety/training equipment	$ 20	$ 20	$ 20	$ 20
	Fees or taxes	None	none	none	none
	Manifests, test reports	$ 15	$ 100	$ 60	$ 10
Insurance		N/A	N/A	N/A	N/A
Production costs	$/Unit of product	$ —	$ —	$ —	$ —
	Labor/Unit Product	$ —	Nominal	Nominal	Nominal
Maintenance	Time	$ 320	$ 860	$ 860	$ 900
	Materials	$ —	$ —	$ —	$ 300

(*Continued*)

TABLE 7.22 *(Continued)*

Operating Costs	Description	Old Process	Ship Waste Off-Site	Discharge to Sewer	Recycle w / UF
Utilities	Water	$ —	$ 200	$ 200	$ 100
	Electricity	$ 250	$ 100	$ 100	$ 300
	Gas/steam	N/A	N/A	N/A	N/A
Annual operating costs		$ 3,472	2,300	$ 3,153	$ 1,680
Cash Flow Summary	**Description**	**Old Process**	**Ship Waste Off-site**	**Discharge to Sewer**	**Recycle w /UF**
Total operating costs		$(3,472)	$(2,300)	$(3,153)	$(1,680)
Incremental cash flow			$ 1,172	$ 319	$ 1,792
−Depreciation			$ (38)	$ (125)	$ (438)
Taxable income			$ 1,135	$ 194	$ 1,355
Income tax (40%)			$ (454)	$ (78)	$ (542)
Net income			$ 681	$ 116	$ 813
+Depreciation			$ 38	$ 125	$ 438
After-tax cash flow			$ 718	$ 241	$ 1,250
Present value	8 yr = 5.7466		$ 4,127	$ 1,387	$ 7,184
Total capital cost			$ (300)	$(1,000)	$(3,500)
Net present value			$ 3,827	$ 387	$ 3,684

ultrafilter and reuse the water. The coolants accounted for 30–60% of the total hazardous waste-disposal costs of the company (oily wastewaters are considered hazardous wastes in this state).

Financial Analysis Conclusions

Having the equipment available to recycle the coolants reduced the need to dispose of fouled coolant. The company conserves water by reusing the filtered water for makeup of new coolant and cleaner. The increased project costs are offset by the decreased need to treat or dispose of wastes.

Unique Features of This Case

This case study underscores the intuitive notion that recycling is profitable whereas treatment is not. Also note that TCA can be used effectively on small projects within small companies as well as large projects in large companies.

CASE STUDY 17: Barrel Electroplating

This small job-shop electroplater is experiencing unexplained plating solution growth. The issue is significant enough that the plating tank must be pumped out every day, resulting in 50 gal of dilute plating solution. Some of this can be used to replenish the bath, but the majority of the solution cannot be reused. The company has requested assistance in determining the cause of the solution growth and evaluating possible remedies.

Step 1: Review of the Production Process

The single line causing the largest problem is the cyanide brass line. This line consists of a hexahedral barrel to hold the parts during plating, 19 tanks, and a load/unload station. The line's hoist is controlled by a tape reader that tells the hoist which tank to visit next, how long to stay submerged, and when to move to the next tank. A simplified process flow is shown in Figure 7.15. Briefly, the line contains a series of cleaning tanks, rinse tanks, an acid activation tank and rinses, brass cyanide tanks and rinses, a sulfuric acid dip, and a nickel plate and rinse tank.

We decide to conduct a time-and-motion study of the hoist travel patterns to determine how long the hoist remains over each tank before proceeding to the next station. Our conclusions are that the hoist travels

■ **FIGURE 7.15 Process Flow Diagram for Barrel Plating Line**

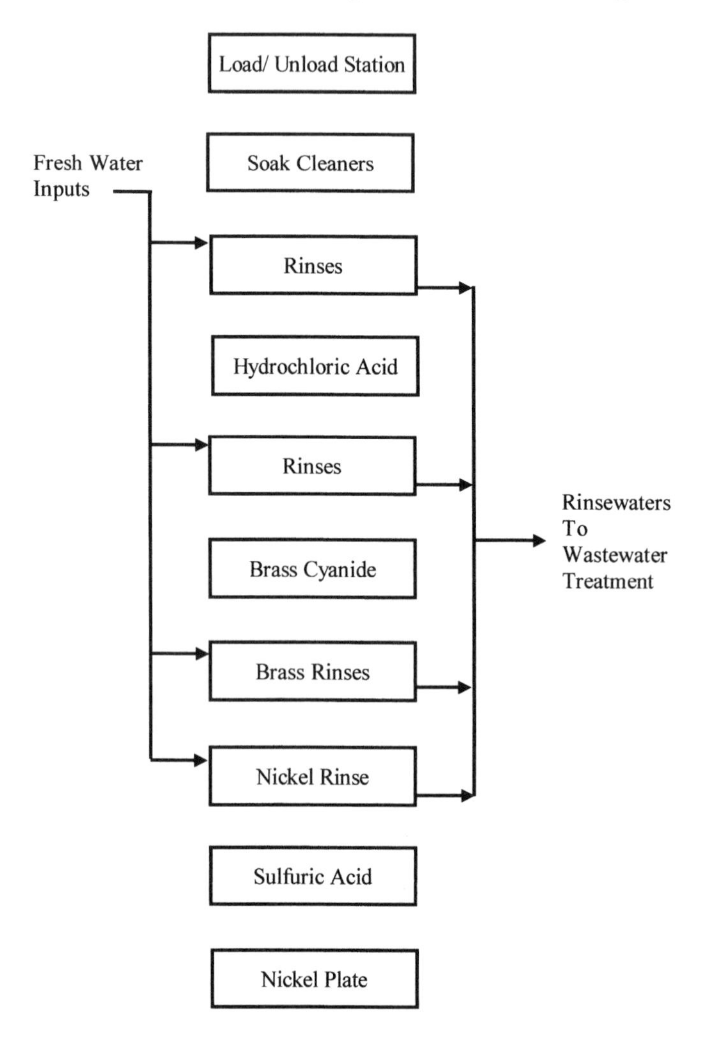

over other tanks before it is fully drained, causing more process solutions to enter the cyanide plating tanks than are carried out.

Step 2: Gathering Cost Data

We obtained the following cost data for the current plating arrangement with the help of the production manager. The company does not have any of the information on computer, although it is building a database. Because the chemical purchase data are not allocated to each plating finish (i.e., nickel, hard chrome, cyanide brass, etc.), we have to estimate the usage according to plating formulation and the proportion of work sent through each line.

Table 7.23 shows the operating costs. Costs are divided by the ten plating lines within the company for all chemicals and inputs that would be distributed evenly. Some chemicals are used mostly in the cyanide plating lines and are not distributed evenly.

Step 3: Defining Alternatives

We have studied the dragout rates and computed hypothetical losses from each major hoist movement that travels across the cyanide brass tanks. We have approximated the total amount of dragout that should be coming off the hoist as 0.31 gal/bbl. This equates to 44.96 gal/day. It therefore seems most likely that the rate of solution dragin to the plating tank is greater than the solution dragout, causing the growth of the solution. The options for minimizing solution gain include:

- Longer hang times over each tank
- Slower withdrawal rates
- Adding a second hoist to reduce hoist travel

The current programming tape for the hoist has the longest possible hang times over each tank. The company must process a given number of barrels per hour to be profitable on this line. Slower withdrawal rates therefore would be out of the question. The most technically feasible option therefore is to purchase a second hoist.

The potential time savings on reduced trips for one cycle is at least half (38 seconds/cycle) of the total of these four trips. A more precise estimate could be developed but is beyond the scope of this project.

Given that each barrel visits up to 14 tanks, the time savings from having a second hoist could be distributed across these 14 visits to decrease process solution dragout. If the withdrawal rate and dwell time over each tank was increased by 1 sec, this would use less than half of the 38-sec time savings achieved.

Extending the dwell times over cleaner tanks, the HCl tank, and each of the brass cyanide rinses would reduce solution losses from 20–35%. This translates directly to chemical savings (approximately 30% for chemicals used on this line) and reduced hazardous waste generation (approximately 30% for waste attributable to this line).

TABLE 7.23 Operating Cost for Plating Line

Cost Item	Description	Cost Factor	Unit Price	Total Units	Total Cost
Purchase costs	Brass solution	100%	$1	45,373	$ 45,373
	Zinc cyanide	20%	$1	1,480	$ 296
	Sodium cyanide (brick)	20%	$1	11,340	$ 2,268
	Sodium cyanide (granule)	80%	$1	5,860	$ 4,688
	Copper cyanide	0%	$1	10,175	$ —
	Potassium cyanide	80%	$1	9,128	$ 7,302
Waste management	Treatment chemicals		$1		$ —
	Sodium hydroxide	10%	$1	19,907	$ 1,991
	Sodium hypochlorite	30%	$1	32,091	$ 9,627
	Miscellaneous chemicals	10%	$1	2,633	$ 263
	Other polluting chemicals	10%	$1	15,938	$ 1,594
	Testing		$1	1	$ —
	Metal hydroxide sludge	10%	$1	59,009	$ 5,901
	Waste cleaners disposal	10%	$1	28,365	$ 2,837
Regulatory compliance	Safety /training equipment			0	$ —
	Water tax	10%	$1	73,000	$ 7,300
	Sewer use fee	10%	$1	5,450	$ 545
	Manifests, test reports	—			$ —
Insurance		—			$ —
Production costs	$/unit of product	—			$ —
	Labor/unit product	—			$ —
	Other	—			$ —
Maintenance	Time	100%	$1	100	$ 100
	Materials	0%		0	$ —
Utilities	Water	10%	$1	38,000	$ 3,800
	Electricity	10%	$1	89,000	$ 8,900
	Gas/steam	0%		0	$ —
Total annual operating cost					$102,785

Step 4: Conducting a Balanced TCA

Cost Assumptions

1. Table 7.24 details the TCA calculations and incremental cash flows. Production costs and potential sales volume were not part of this calculation. Equipment vendors claim a 80–100% increase in throughput capacity of loads/hour. The slower withdrawal and extended dwell time will allow the company to maintain compliance with wastewater discharge limits and increase the number of barrel loads per hour on this line.
2. Electricity costs are based upon annual facility usage figures. The total cost was divided evenly among the nine plating lines and the other plant operations (10% of $189,000 was allocated to the brass line). The cost appearing in the financial analysis ($6,300) is an estimation of the electricity used by the existing hoist on the brass line (one third of $18,900). Adding a second hoist was assumed to be less than twice the current cost, as some savings would be accrued from the hoist not traveling as much ($6,300 + 80% = $11,340).
3. Reductions in process chemical dragout achieved through slower withdrawal time and extended dwell time would decrease brass and other cyanide chemical consumption by 30%. Nickel usage was not included in this analysis, but would see similar reductions.
4. Waste treatment and disposal costs associated with the brass line would also decrease 30–50% due to reductions of process chemical drag-in into running rinses.
5. The drag-out reductions would precipitate decreases in water use, wastewater flow discharge and fees charged for sewer usage and water taxes. Conservatively, this savings was given as 20% to adjust for rate increases.

Financial Analysis Conclusions

This analysis uses a 10-year lifetime for the new hoist, depreciated on a straight-line basis. Adding a second hoist has a net present value of $16,244 over the project's 10-year life. That is, the project will pay for itself and yield a profit of $16,244. Using the costs listed above the project pays back in 3.2 years.

The calculations in the table did not address the potential impact the second hoist would have on production capacity. If production

■ TABLE 7.24 Financial Analysis of the P2 Project

Capital Costs	Description	Current Process	Second Hoist
Equipment purchase			$ 35,000
Disposal of old process			$ —
Research & design			$ —
Initial permits			$ —
Building/process changes			$ 5,000
Total capital costs			$ 40,000
Operating Costs	**Description**	**Current Process**	**Second Hoist**
Purchase costs	Brass solution	$ 4,537	$ 3,175
	Ancillary	$14,554	$ 10,187
Waste management	Treatment chemicals	$13,475	$ 9,027
	Testing	N/A	N/A
	Metal hydroxide disposal	$ 5,901	$ 4,131
	Waste cleaner disposal	$ 2,837	$ 1,985
Regulatory compliance	Safety/training equipment	$ —	$ —
	Fees or taxes	$ 7,845	$ 6,275
	Manifests, test reports	$ —	$ —
Insurance		N/A	N/A
Production Costs	$/Unit of product	$ —	$ —
	Labor/unit product	$ —	$ —
Maintenance	Time	$ 100	$ 200
	Materials	$ —	$ —
Utilities	Water	$ 3,800	$ 3,040
	Electricity	$ 8,900	$ 11,340
	Gas/steam	$ —	$ —
Annual operating costs		61,949	$ 49,360

(Continued)

■ **TABLE 7.24** ***(Continued)***

Cash Flow Summary	Description	Current Process	Second Hoist
Total operating costs		$(61,949)	$(49,360)
Incremental cash flow			$ 12,589
−Depreciation			$ (4,000)
Taxable income			$ 8,589
Income tax (40%)			$ (3,436)
Net income			$ 5,153
+Depreciation			$ 4,000
After-Tax cash flow			$ 9,153
Present value	10 yr		$ 56,244
Total capital cost			$(40,000)
Net present value			$ 16,244

capacity increases were included in this calculation the project would have a net present value of $1.1 million and a payback of 1.5 months!

Unique Features of This Case

This case illustrates the depth that is sometimes required to reveal answers to a problem. In this instance, the second hoist would not have been an obvious answer to the problem of solution growth. Only through studying the process both growth were we able to see the first hoist was traveling back over the plating tanks too often.

CASE STUDY 18: Closed-Loop Electroplating

This case study will compare cost scenarios under no pollution control, exclusively pollution treatment, and under pollution prevention for a hard chrome electroplating line. Specifically, the first cost scenario will be the cost of installing and operating a new hard chrome electroplating line. The second will be the addition of the required air handling equipment (mesh pad filters) and counterflow rinsing system to meet environmental requirements. The third will be the reuse of the mesh pad washwaters and dragout solution.

Step 1: Review of the Production Process

This small electroplating facility performs custom plating and metal finishing of pre-manufactured parts. The company provides nickel, decorative chromium and hard chromium electroplated finishes and engineer-customized rack arrangements for unique part shapes.

The company expanded its operations in 1994 by installing four new hard chrome plating tanks. Compliance with environmental regulations was made a priority during construction of the new facility. Environmental design elements included setting the tanks over a lined, sealed holding sump to retain any spills. The building floor was poured as one continuous form, 6 inches below the foundation walls. Between the floor, the foundation, and the sump walls and sump floor, a 2-inch Bentonite seal prevents liquids from seeping through these joints. See Figure 7.16.

Step 2: Gathering Cost Data

The company president was directly involved in gathering the cost data, which ensured both accuracy and timeliness. The company's purchase records and hazardous materials are all on a computer data base enabling the president to retrieve several different reports showing usage over time and annual costs.

Most of the labor for this project was contractor labor and was included in the capital cost as part of the total purchase price for the

■ **FIGURE 7.16 Process Flow Diagram for Chrome Plating Line**

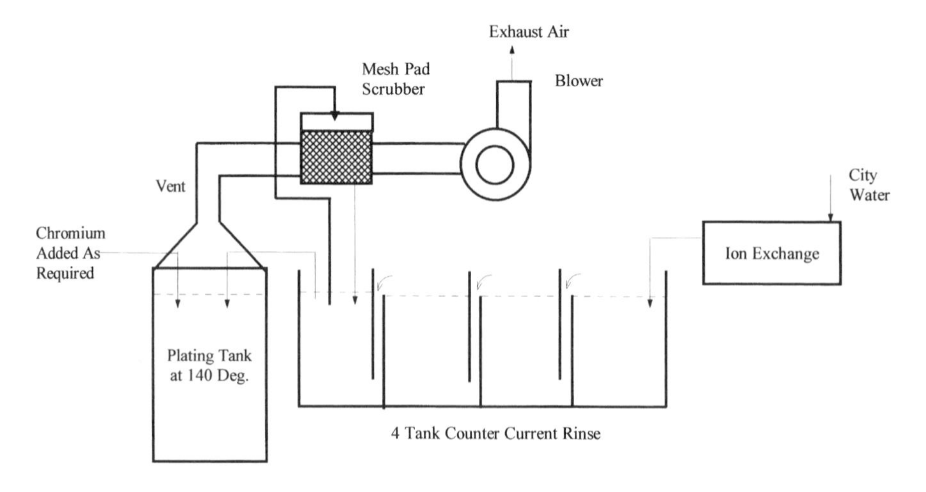

equipment. The number of hours spent researching and designing this system was not tracked. The plating lines and air handling equipment were installed in a new structure which did not require the movement of walls or equipment to install the plating lines.

Steps 3 and 4: The Alternatives and the TCA

For the sake of clarity, the steps of describing each alternative and then the cost assumptions used for the TCA have been combined. Each larger section describes the equipment installed, the costs associated and any savings and impacts on performance of the plating line.

Traditional Plating Scenario The traditional four-stage running rinse without air handling equipment represents plating operations before the entry of air pollution control regulations. The operating costs for this scenario are the largest of the three owing to the lack of water-conserving practices. Operating costs for the new line follow typical chemical usage and maintenance expenditures. Four running rinses fed by city water enter the rinse tanks and exit to the wastewater treatment system. Deionized water as a rinsewater source would be cost prohibitive in this scenario.

Four Stage Countercurrent Rinse Scenario This scenario changes the running rinses to a four-stage countercurrent rinsing arrangement, reducing water usage by eliminating three running rinses and wastewater treatment maintenance, while decreasing the need to replace treatment system parts. The change saves over $9,000 per year (after taxes). The small amount of plumbing needed to accomplish the counterflow arrangement more than pays for itself.

Federal environmental regulations require hexavalent chromium plating tanks to have mechanical ventilation and air pollution control equipment to treat ventilated plating fumes. The company's system filters air through mesh pads that trap chrome containing droplets. These pads are periodically sprayed down with water to clean off the chromium solution, which is then sent to the wastewater treatment system.

The additional capital costs for air pollution control equipment are shown in Table 7.25 and total $11,000. This cost includes the ductwork, blowers and mesh pad filters for removing chromium vapor from the

■ TABLE 7.25 Capital Costs for Each Plating Arrangement

	Traditional Plating Arrangement	Countercurrent Rinse + Air Equipment	Closed Loop Process
Equipment purchase			
Plating tank	(2,500)	+	+
Rectifier	(15,000)	+	+
Anodes	(4,800)	+	+
Bussing	(2,000)	+	+
Solutions	(2,500)	+	+
Steam/cooling	(4,000)	+	+
Fittings	(2,000)	+	+
Hoist system	(2,500)	+	+
S.S. rinse tanks	(8,000)	+	+
Blower	—	(3,500)	+
Scrubber (4 stage)	—	(2,500)	+
Hood	—	(1,500)	+
Ductwork	—	(2,000)	+
Air pump	—	—	(800)
Filter	—	—	(600)
Piping	—	—	(600)
Disposal of old process	N/A	N/A	N/A
Research & design	not tracked	0	0
Initial permits	none required	0	0
Contracted labor	(8,000)	(1,500)	(500)
Building/process changes	none	none	none
Total capital costs	(51,300)	(62,300)	(64,800)
Incremental capital costs	N/A	(11,000)	(13,500)

airstream. The installation of air scrubbers adds a $2,420 labor cost for washing down the mesh pads and reassembly of the scrubbers.

Closed-Loop Scenario The company repiped the mesh pad filter washwaters into the first plating rinse tank. Rerouting the mesh pad washwaters back into the dragout tank captures 90–95% of the chromium solution trapped in the mesh pads and recycles it back to the plating process. This tank has no outlet, serving as a still rinse or "drag-out" tank. Fresh water inflow to the other rinses was reduced to an on-need basis to replenish evaporative losses and made the use of de-ionized water affordable.

The four closed loop chromium tanks contribute only 100 gallons per day in floor washings to the 25,000 gallons per day wastewater flow. These daily floor washings are treated along with rinsewaters from other processes. The four plating tanks share a cleaning line whose 2-gallon per minute rinses are sent to the wastewater treatment system. The floor washings and the parts cleaner rinse waters are the closed loop system's only contribution to wastewater discharges (estimated to be 9% of the total facility water flow). The staff uses a fog wand for rinsing the rack as it rises out of the plating tank. This further reduces drag-out.

The chrome-lead sludge and other impurities build up on the bottom of the tank which must be pumped out and sent off-site for recycling. The company has not performed this task yet, but estimates the tanks will only need pumping and replacement every three years instead of annually.

Step 4: Conducting a Balanced TCA

Table 7.26 details the TCA calculations and incremental cash flows.

Cost Assumptions

1. Water usage in the non-counterflowed rinses are based on typical operating practices on other lines at the facility. A 2-gallon per minute flow would be used on the noncounterflowed rinses and the on counterflowed rinse of the second scenario. Freshwater in-flow in the closed loop scenario is dictated by evaporation rates of the four chrome process tanks. This flow averages 0.18 gallons per minute over a typical operating day.
2. Ion exchange would have been cost prohibitive with four, 2-gallon per minute running rinses due to the extraordinary expense of treating that amount of incoming, once-through water (approximately 2 million gallons per year).
3. The labor requirements for the wastewater treatment system were estimated based upon the increased flow to the system and the larger proportion of that flow coming from the non-counterflowed rinsing system.

■ TABLE 7.26 Financial Analysis of P2 Project

Capital Costs	Description	Traditional Plating Line	CCR +Air Equipment	Closed Loop
Plating equipment		$43,300	$43,300	$43,300
Contracted labor		$ 8,000	$ 9,500	$10,000
Air pollution control equipment		$ —	$ 9,500	$9,500
Closed loop equipment		$ —	$ —	$2,000
Building/ process changes		$ —	$ —	$—
Total capital costs		$51,300	$62,300	$64,800
Capital Costs	**Description**	**Traditional Plating Line**	**CCR +Air Equipment**	**Closed Loop**
Chemical purchases	**Process bath adds**	$12,250	$12,250	$12,140
	Ancillary chemical adds	$ 660	$ 660	$600
	Bath make-up (every 3 yr)	$ 2,920	$ 2,920	$2,920
	Process bath testing	$ 400	$ 400	$400
Waste management	Treatment chemicals	$ 3,190	$ 1,730	$970
	Tank pump outs	$ 1,750	$ 1,750	$1,750
	DMR testing	$ 140	$ 140	$110
	Disposal	$ 2,930	$ 1,250	$680
Regulatory compliance	Safety/training equipment	$ 30	$ 30	$30
	Hazardous waste tax	$ 100	$ 40	$20
	Manifests, test reports	$ 100	$ 100	$100

(Continued)

■ **TABLE 7.26** *(Continued)*

Capital Costs	Description	Traditional Plating Line	CCR +Air Equipment	Closed Loop
Maintenance	Compressor & pump labor	$ —	$ —	$280
	Compressor & Pump materials	$ —	$ —	$ 220
	Air scrubber labor	$ —	$ 2,420	$ 2,420
	Air scrubber materials	$ —	$ —	$ —
	WWT system labor	$ 14,900	$ 6,800	$ 3,690
	WWT system materials	$ 2,520	$ 2,100	$ 1,890
Utilities	Water	$, 9,120	$ 4,950	$ 2,920
	DI column rental	$ —	$ —	$ 4,260
	Electricity	$ —	$ —	$ 230
	Gas/steam	not tracked	not tracked	not tracked
Annual operating costs		51,010	$ 37,540	$ 35,630
Cash Flow Summary	**Description**	**Traditional Plating Line**	**CCR +Air Equipment**	**Closed Loop**
Total operating costs		$(51,010)	$(37,540)	$(35,630)
Incremental cash flow			$ 13,470	$ 15,380
−Depreciation			$ (6,230)	$ (6,480)
Taxable income			$ 7,240	$ 8,900
Income tax (40%)			$ (2,896)	$ (3,560)
Net income			$ 4,344	$ 5,340
+Depreciation			$ 6,230	$ 6,480
After-tax cash flow			$ 10,574	$ 11,820
Present value	10 yr		$ 64,973	$ 72,629
Total capital cost			$(62,300)	$(64,800)
Net present value			$ 2,673	$ 7,829

Financial Analysis Conclusions

The company tracked a variety of costs connected to the development of the closed loop system. All large capital costs including rectifiers, plating anodes, tanks, hoists, blowers, ductwork and filters were accounted for. Building the plating lines represents 79% of all costs for the three scenarios combined. The air handling equipment cost an additional $11,000 and accounted for 17% of the total cost. Closing the loop cost $2,500, or the remaining 4% of the project cost.

Three comparisons of financial viability are detailed below that summarize the costs and benefits of each scenario.

1. The air equipment accounts for almost all non-process capital costs, but it generates no cost savings in this project. Rather it increases operating costs through additional labor required to clean and maintain the scrubber mesh pads. The savings from counterflowing the rinses (in water use, treatment chemicals and treatment system maintenance labor) balances out the increased labor cost for maintaining the air scrubber system. The size of these savings also makes the project look desirable with the capital cost recovered in just over 22 months. This project's value is 4.76 times the initial investment (present value/total capital cost).
2. The capital cost for installing the air equipment, countercurrent rinses, and pumps and filters to close the loop totals $13,500. This investment yields an annual cost savings of $7,180 totaling over $30,000 in the 10-year lifetime of the project. More simply put, this project's value is 4.44 times that of the initial investment.
3. The capital cost for just closing the loop is $2,500. The benefits directly attributable to this action total $1,240 per year. The project's present value is 3.05 times that of the initial investment. The savings are not as dramatic as the other scenarios, but represent tangible benefits available at relatively low levels of capital investment. Additional benefits include smaller quantities of metal hydroxide sludge, conservation of process metals and re-use of process rinsewaters. Closing the loop minimizes water use to a level where de-ionized water becomes economical. This results in better rinsing and fewer bath contaminants.[7]

[7]The author gratefully acknowledges ConnTAP, Hartford, Connecticut, and GZA GeoEnvironmental, Inc., Vernon, Connecticut, for their assistance with this case study.

Unique Features of This Case

One of the most interesting aspects of this case study is the comparison available between plating without any environmental controls, plating with just traditional waste treatment and air pollution control, and then plating with pollution prevention technologies in place. Here is verifiable proof that pollution prevention makes a difference from the beginning.

CHAPTER 8

PRACTICE EXERCISES

The only way to thoroughly understand the TCA process is through direct experience and practice. The following exercises will help you become familiar with Total Cost Assessment techniques. The exercises are based on pollution prevention projects at real manufacturing facilities, but were adapted to highlight difficulties encountered in the field. Each one emphasizes particular aspects of the TCA process or a situation requiring some creativity.

HOW TO USE PRACTICE EXERCISES

The exercises progress from the most simple to the most complex by introducing additional options to be evaluated, more complex data, and options with varying lifetimes. These exercises can be worked in several ways. For those who still need more information on the mathematical portion of the TCA process, you may wish to look ahead at the answers and try to determine how the calculations were performed. Those who are fairly comfortable with the mechanics but shaky on the data-gathering portion should spend more time reading the text and extracting the required data. Each exercise has a section of notes at the end explaining the unique aspects of the exercise, the observations and assumptions made, and some of the difficulties encountered.

ADAPTING FOR CLASSROOM TEACHING

These exercises have been used in conference presentations, workshops, and course curricula. To adapt them for classroom use, consider the level of understanding and abilities of the students. Keep in mind that these exercises require a significant amount of time to work through and to review the answers.

The exercises can be given as group activities, with each member of the group playing a particular role in the assessment process. For example, information in Exercise 2 is provided in categories based on factory staff that would be part of a pollution prevention team (purchasing, environmental management, engineering, and manufacturing). Each person in a group could adopt the role of one of these team members for this exercise. Each team member has information necessary for an accurate assessment of the project, requiring all members to participate. Omission of any of the pieces results in an incorrect answer. This method also helps students develop team working skills.

Exercise 1 uses group discussion to identify costs involved in a TCA. The emphasis is on understanding what data elements are required to conduct the TCA and where you would look to find the information or how it would be estimated.

Exercise 2 is a case study involving an electroplating company desiring to replace a cyanide-based plating chemical with a less toxic alternative. The shortcut TCA is used, but enough data are presented to conduct a balanced TCA if desired.

Exercise 3 is an intermediate-level case study involving a textile company searching for a replacement for copper-based dyes. This exercise uses the balanced TCA and presents the common conflict of desiring to pursue pollution prevention and the necessity of achieving compliance in a short timeline.

Exercise 4 is a case study about alternatives to solvent-based spray paints. This is the most complex of the four practice exercises and can be performed either as a balanced TCA with the provided worksheets or as an advanced TCA if you developed your own spreadsheets.

EXERCISE 1: Identifying Costs

Instructions for Exercise 1

You are the pollution prevention team leader for a manufacturing plant. Your team has decided that replacing the current vapor degreaser with a new technology will be the next pollution prevention project.

1. What basic cost categories would the team need to consider in an assessment of financial feasibility?
2. Where would the team find cost data for these categories?
3. What additional factors might influence the feasibility of this project?

Answers for Exercise 1

1. Review Chapter 2: Gathering Cost Data. Costs to be considered include process and ancillary chemical purchases; storage costs; waste testing; treatment and disposal; worker health and safety training; insurance; chemical, water and sewer usage fees; interest on loans for the project; labor associated with maintenance and paperwork; production costs; utilities; research and design; initial and annual permits; disposal of old process; changes to the building or processes; retraining workers; potential loss of jobs; and purchase of the new equipment.
2. Sources for information include the purchasing department; line workers; process engineers; accountants; vendors; Environmental, Health & Safety Environmental, Health & Safety (EH&S) staff; insurance agent; waste manifests; product records; chemical formulations; material safety data sheets; and utility bills.
3. Additional factors that might infuence feasibility include new permits for wastes associated with new process; taxes and fees associated with new chemicals; purchase costs of equipment and useful life of new technology; new or proposed legislation that would have a bearing on the new technology; and work health and safety issues surrounding the new technology (i.e., toxicity, carcinogenicity, ventilation issues, explosion risks).

EXERCISE 2: Electroplating Analysis

Instructions for Exercise 2

Use the process and information developed in Exercise 1 and the following proposals to assess the feasibility of the following pollution prevention project. Incorporate both quantitative and qualitative analyses. For the shortcut TCA, select the largest operating cost categories and only the major purchases for the capital costs. Jot these down on scrap paper and try to calculate the NPV without a calculator. If you are performing the balanced TCA, use the worksheets provided on the following pages. Answers to both approaches can be found on pages 251–255.

1. Use the provided role information to fill in the table of costs. Calculate capital costs for each alternative, the total operating cash flows, and the incremental savings.
2. Calculate the present value (PV), the net present value (NPV), and the ratio of benefits to costs. Is this alternative feasible?

American Electro-Metals, Inc.

American Electro-Metals, Inc. (AEM), is a medium-size, privately owned, high-volume, electroplating job shop. AEM plates a variety of finishes: nickel, bright brass, cadmium, and chromates. Most plating equipment is 10–20 years old. AEM's second largest waste stream is excess cyanide brass plating solution. This excess is pumped out, placed in barrels, and the bath reformulated resulting in production downtime. AEM accumulates dilute brass cyanide solution at a rate of 44 gal/day. The company could be cited for "treating" hazardous waste illegally if it attempts to empty the barrels into its wastewater treatment system, even though it is permitted for treating cyanide rinsewaters. Currently, the extra solution is hauled off-site as hazardous waste.

AEM's quality/environmental team has been instructed to find a way to decrease the costs associated with hauling the waste off-site. They think the cause of the problem is too much dragin from previous tanks and too little dragout from the brass bath. Options considered were hard-piping the cyanide tank to the treatment system and obtaining a permit modification to treat process solutions; low-temperature evaporation of the excess solutions and possible reuse; adding a second hoist to the plating line to decrease dragin of other process solutions and rinsewaters. The team decided to evaluate adding the second hoist because of the potential benefits to product quality and increases in production capacity. The company's discount rate is 10%, with a 40% tax rate.

Adding a Second Hoist

The group conducted a time-and-motion study of hoist dwell and travel patterns and concluded that extending dwell times could reduce solution losses from 20–35%. This translates directly to chemical savings (approximately 30% for chemicals used on this line) and reduced hazardous-waste generation.

The group agreed that the new hoist would have automated indexing and computer timing features to control withdrawal rates from each tank and dwell times over the tanks, and also would be compatible with the existing hoist. Jessup Plating Equipment, Inc., provided AEM with electrical, space, and weight requirements for this second hoist. The proposal included labor for reprogramming the first hoist, and testing and fine-tuning both hoists to work in tandem. Conservatively, the hoist was given a 10-year lifetime. Additional information includes the following.

Purchasing

- The cost estimate for the second hoist comes to $35,000.
- Reductions in process chemical dragout achieved through slower withdrawal time and extended dwell time would decrease brass and other cyanide chemicals consumption by 30%.

Environmental Management

- Waste-treatment chemicals and disposal costs associated with the brass line would decrease 33% due to reductions of process chemical dragin into running rinses.
- The costs for filling out manifests, test reports, conducting personnel training, or insurance would not be effected by this process change; therefore, no costs were developed.

Engineering

- Maintenance costs for this line would double with the addition of a second hoist. An estimated $5,000 of building alterations to support the second hoist would be needed.
- For the initial project analysis, the costs of production (increases or decreases) will not be considered.

Manufacturing

- Adding a second hoist is expected to be less than twice the current electricity cost, as some savings would be accrued from the hoist not traveling as much ($6,300 + 80% = $11,340).
- The dragout reductions would decrease water use, wastewater flow discharge, sewer usage fees, and water taxes charged to AEM. Conservatively, water usage and the associated sewer usage fee was given a 20% decrease to adjust for rate increases. Water tax was given a 10% decrease.

Answers for Exercise 2: Shortcut Approach

See Tables 8.1 to 8.3.

■ TABLE 8.1 Capital Cost Worksheet

Capital Costs	Current Process	Second Hoist
Equipment purchase	N/A	
Disposal of old process	N/A	0
Research & design	N/A	0
Initial permits	N/A	0
Building/process changes	N/A	
Total capital costs	N/A	

■ TABLE 8.2 Operating Cash Flow Worksheet

Operating Cash Flow		Current Process	Second Hoist
Chemical purchases		($19,091)	
Waste management	Chemicals	($13,473)	
	Testing	N/A	
	Disposal	($8,736)	
Safety training/equipment		N/A	
Insurance		N/A	N/A
Sewer use fees		($545)	
Filing paperwork time		N/A	N/A
Water tax		($7,300)	
Production costs	% inc./dec.	(?)	(?)
	$/yr	(?)	(?)
Maintenance	Time	($100)	
	Materials	0	0
Utilities	Water	($3,800)	
	Electricity	($6,300)	
	Gas/steam	0	
Total annual Operating Cash Flow			

Largest Capital Costs

1. Second hoist ($35,000)

Top Three Operating Cash Flows	***(Before)***	***(After)***	***(Difference)***
1. Chemical purchases	($19,091)	($13,363)	$5,728
2. Waste chemicals	($13,473)	($9,027)	$4,446
3. Waste disposal	($8,736)	($5,853)	$2,883

■ **TABLE 8.3 Cash Flow Summary Worksheet**

Cash Flow Summary	Current Process	Second Hoist
Total operating costs		
Incremental cash flow		
−Depreciation		
Taxable income		
Income tax (40%)		
Net income		
+Depreciation		
After-tax cash flow		
Present value		
Total capital cost		
Net present value		
Cost–benefit ratio		

Total Operating Cash Flow	***(Before)***	***(After)***	***(Difference)***
	($41,300)	($28,243)	$13,057
Annual savings =			$13,057
× present value factor for 10 years at 10%			6.1446
Present value of second hoist project			$80,230
Shortcut approach net present value			$45,230

Answers for Exercise 2: Balanced Approach

See Tables 8.4 to 8.6.

Financial Feasibility

Shortcut Approach The quick analysis portrays the project to be very profitable. This turns out to be rather misleading on closer examination. There are two reasons for this discrepancy:

1. The capital cost has a second cost for building modifications that represents 12% of the total project cost. This forces the project to have a higher cost savings in order to have a positive net value.
2. Other operating costs would appear to be significant in this case, such as the increase in utility costs that was overlooked in the shortcut approach because they are the fourth largest and only the top three costs were examined. This has the effect of decreasing the overall cost savings, and the present value.

■ **TABLE 8.4 Answer Key 1**

Capital Costs	Current Process	Second Hoist
Equipment purchase	N/A	($35,000)
Disposal of old process	N/A	0
Research & design	N/A	0
Initial permits	NA	0
Building/ProcessChanges	N/A	($5,000)
Total capital costs	N/A	($40,000)

■ **TABLE 8.5 Answer Sheet 2**

Operating Cash Flow		Current Process	Second Hoist
Chemical purchases		($19,091)	($13,363)
Waste management	Chemicals	($13,473)	($9,027)
	Testing	N/A	N/A
	Disposal	($8,736)	($5,853)
Safety training/equipment		N/A	N/A
Insurance		N/A	N/A
Sewer use fees		($545)	($436)
Filing paperwork time		N/A	N/A
Water tax		($7,300)	($6,570)
Production costs	% inc./dec.	(?)	(?)
	$/yr	(?)	(?)
Maintenance	Time	($100)	($200)
	Materials	0	0
Utilities	Water	($3,800)	(3,040)
	Electricity	($6,300)	($11,340)
	Gas/steam	0	0
Total annual operating Cash Flow		($59,345)	($49,829)

Balanced Approach This analysis used a 10-year lifetime for the new hoist, depreciated on a straight-line basis. The corporate tax rate was assumed to be 40% and the hurdle rate for this investment was set at 10%. This results in a net present value of $4,917 over the project's 10-year life. That is, the project will pay for itself and yield a profit of $4,917. By using only the previously listed costs, the project pays back in 4.2 yr.

■ **TABLE 8.6 Answer Key 3**

Cash Flow Summary	Current Process	Second Hoist
Total operating costs	($59,345)	($49,829)
Incremental cash flow	N/A	$9,516
−Depreciation	N/A	($4,000)
Taxable income	N/A	$5,516
Income tax (40%)	N/A	($2,206)
Net income	N/A	$3,310
+Depreciation	N/A	$4,000
After-tax cash flow	N/A	$7,310
Present value	N/A	$44,917
Total capital cost	N/A	($40,000)
Net present value	N/A	$4,917

Teacher's Notes for Exercise 2

What would be the revised net present value, given the following assumptions about the impacts on production costs?

- A unit cost to AEM's customer of $18/bbl of brass plated parts.
- An increase in production capacity from 7 to 12 bb/hr.
- A 4,000-hr work year.

The answer to how much production costs would affect the analysis is: The net present value would rise to $1.3 million, and the project would pay for itself in 1.3 months.

- Adding the second hoist could potentially double production capacity on the cyanide brass line.

What is a logical argument against including production costs?

- You may have the production capacity to run twice as much product, but you may not always have the demand.

What would be the likely net present values for the other proposed projects (hard-piping waste solutions to the wastewater treatment system and installing a low-temperature evaporator)?

- For the hard-pipe option, there would be lower capital costs than installing a second hoist, except for the permitting process. There would be fewer savings in operating costs, with the decrease in hazardous-waste costs offset by the increased use of wastewater treatment chemicals and testing. Other impacts?

- For the low-temperature evaporator, there would be the capital cost of equipment purchase and installation. Utilities use would increase (electricity); process chemical usage and waste disposal costs would decrease. Other impacts?

EXERCISE 3: Textile Dyeing Analysis

Instructions for Exercise 3

This exercise is similar to Exercise 2 and provides another real-life scenario. Again, use the process and cost information to assess the feasibility of the following pollution prevention projects. Incorporate both quantitative and qualitative analyses. Use the balanced approach and the worksheets provided on the following pages. Answers can be found on pages 000 and 000.

1. Use the provided role information to fill in the table. Calculate capital costs of each alternative, total operating costs, and incremental cash flow.
2. Calculate the present value (PV) and the net present value (NPV) for each alternative.
3. Answer the following questions:
   a. At which points in the analysis does the attractiveness of the different options change?
   b. Which alternatives can be eliminated based on their net present value?

Tightwove Fabrics, Inc.

Tightwove Fabrics, Inc., is a small-size, privately owned, textile dye shop providing dyed and finished fabrics to a variety of sportswear and garment companies. Most dyeing equipment is 20–30 years old. Tightwove is located within an urban center and has very little extra floor space. The company is seeking a pollution prevention alternative to meet a new discharge limit for copper.

The team has been instructed to look at technologies for reducing or eliminating copper in the wastewater. Review of the processes, material safety data sheets, and dye bath recipes revealed two options: Find a substitute for the copper dyes or improve the dye/fabric adhesion efficiency. The team located vendors of chemical dyes and dying technologies and received these proposals. The company's cost of capital is 17%.

Option 1: The Blue Sky Dye Proposal

Blue Sky Chemical Company provided the team with MSDSs and samples of copper-free dyes, as well as a list of references. The dyes have no copper-based pigments, but use a zircon-based pigment. The literature said the new pigment was extremely lightfast and colorfast. Color swatches included in the sales material compared favorably with the color hues used at Tightwove. A higher operating temperature and the need for a hot rinse would require some modifications to the existing dye jigs. The sales rep said the EPA has been so busy with solvents and CFCs that the evaluation of zircon's potential toxicity will take at least 4 to 5 years. Additional information includes the following.

Purchasing

- Blue Sky chemicals are 11% more expensive.
- Modifications to the heating lines and piping for new hot rinses will cost $31,000.

Environmental Management

- The lack of effluent limits for zircon will decrease the need to install a wastewater treatment system.
- The lack of toxicity information on this product raises concerns about worker acute exposure and chronic health effects.
- Blue Sky has a 7-year lifetime due to potential changes in regulations.

Engineering

- Increased temperatures in the tanks and feed lines may fatigue an already well-used system and could precipitate a rupture and spill.
- Maintenance costs will rise on the boiler that provides process heat ($3,000 in labor and $5,000 in parts).
- Other facilities using this chemical have found it difficult to reuse the rinsewaters and maintain product quality.

Manufacturing

- Processing time will decrease 5% due to shorter drying time and will save $6,250.
- The dying process would remain unchanged and not require retraining workers.

- Water use will decrease by 5%.
- Natural gas costs would rise by $2,000.

Option 2: The Speed Jet Proposal

Speed Jet is a manufacturer of ultrasonic assisted dye jigs. The company claims its technology can achieve a 99.5% exhaustion and fixing of dye molecules to fabric, even after rinsing. Typically, exhaustion and adhesion are anywhere from 80–90% in a beck-type operation. The Speed Jet operation would require considerable modifications to the process line. The entire process is one long unit, complete with an infrared drying system and recirculating rinse and filter system. The vendor claims product quality improves because the ultrasonics force the dye into the fibers, resulting in a richer hue. This process was given a 7-year lifetime. Additional information includes the following.

Purchasing

- The machinery will cost $257,000 and building alterations $110,000.
- Tightwove can continue to use the existing dye and would actually use 40% less.

Environmental Management

- There appears to be limited risk to workers from the self-contained units.
- It will cost $1,000 to dispose of the old system.

Engineering

- Processing a variety of jobs with this unit may raise scheduling issues and require more careful monitoring of individual runs.
- Electrical use will rise to $10,000/yr.

Manufacturing

- Workers are concerned that some will lose their jobs. There is skepticism that this technology can work. It looks too delicate.
- Maintenance on this system is estimated at $1,000/yr in labor and $767/yr in parts, and may require out-of-shop specialists.

- Production costs would decrease through faster processing time, providing a 6% annual cost savings.
- All heat for baths and drying is electrical; no steam is needed.
- Water use would decrease by 27% due to a more efficient system.

See Tables 8.7 to 8.9.

TABLE 8.7 Capital Cost Worksheet

Capital Costs	Current Process	Blue Sky	Speed Jet
Equipment purchase	N/A		
Disposal of old process	N/A		
Research & design	N/A	0	0
Initial permits	N/A	0	0
Building/process changes	N/A		
Total capital costs	N/A		

TABLE 8.8 Operating Cash Flow Worksheet

Operating Cash Flow		Current Process	Blue Sky	Speed Jet
Chemical purchases		($87,000)		
Waste management	Chemicals	0	0	0
	Testing	0	0	0
	Disposal	0		0
Safety training/equipment		($25)		
Insurance		N/A	N/A	N/A
Chemical use fees		0	0	0
Filing paperwork time		($100)	($100)	($100)
Annual permitting		($200)	0	($100)
Production costs	% inc./dec.	0	5%↓	6%↓
	$/yr	($125,000)		
Maintenance	Time	($1,000)		
	Materials	0		0
Utilities	Water	($275,000)		
	Electricity	($2,000)	($2,000)	
	Gas/Steam	($14,000)		($8,000)
Total annual operating cash flow				

■ **TABLE 8.9 Cash Flow Summary Worksheet**

Cash Flow Summary	Current Process	Blue Sky	Speed Jet
Total operating costs			
Incremental cash flow			
−Depreciation			
Taxable income			
Income tax (40%)			
Net income			
+Depreciation			
After-tax cash flow			
Present value			
Total capital cost			
Net present value			

Answers for Exercise 3

See Tables 8.10 to 8.12.

Financial Feasibility

Neither option is financially feasible based on their negative net present values. Notice that the annual savings before taxes (incremental cash flow) from the Speed Jet option are significant, but are offset over the project's lifetime by the large capital cost. Also notice that even though the Blue Sky alternative saves $1,600/yr over the life of the project, it will cost $31,188.

Teacher's Notes for Exercise 3

Ask the class to explain what other factors should be included in the analysis of these alternatives.

■ **TABLE 8.10 Answer Key 1**

Capital Costs	Current Process	Blue Sky	Speed Jet
Equipment purchase	N/A	0	($257,000)
Disposal of old process	N/A	0	($1,000)
Research & design	N/A	0	0
Initial permits	N/A	0	0
Building/process changes	N/A	($31,000)	($110,000)
Total capital costs	N/A	($31,000)	($368,000)

■ **TABLE 8.11 Answer Key 2**

Operating Cash Flow		Current Process	Blue Sky	Speed Jet
Chemical purchases		($87,000)	($96,570)	($52,200)
Waste management	Chemicals	0	0	0
	Testing	0	0	0
	Disposal	0	0	0
Safety training/equipment		($25)	($25)	($25)
Insurance		N/A	N/A	N/A
Chemical use fees		0	0	0
Filing paperwork time		($100)	($100)	($100)
Annual permitting		($200)	0	($100)
Production costs	% inc./dec.	0	5%↓	6%↓
	$/yr	($125,000)	($118,750)	($117,500)
Maintenance	Time	($1,000)	($3,000)	($1,000)
	Materials	0	($5,000)	($767)
Utilities	Water	($275,000)	($261,250)	($200,750)
	Electricity	($2,000)	($2,000)	($10,000)
	Gas/steam	($14,000)	($16,000)	($8,000)
Total annual operating cash flow		($504,325)	($502,695)	($390,442)

■ **TABLE 8.12 Answer Key 3**

Cash Flow Summary	Current Process	Blue Sky	Speed Jet
Total operating costs	($504,325)	($502,695)	($390,442)
Incremental cash flow	N/A	$1,630	$113,883
−Depreciation	N/A	($4,428)	($52,571)
Taxable income	N/A	($2,798)	$61,312
Income tax (40%)	N/A	($1,119)	($24,525)
Net Income	N/A	($4,476)	$36,787
+Depreciation	N/A	$4,428	$52,571
After-tax cash flow	N/A	($48)	$89,358
Present value	N/A	($188)	$350,500
Total capital cost	N/A	($31,000)	($368,000)
Net present value	N/A	($31,188)	($17,500)

- What aspects of the analysis were misleading?
- Did people think one option was clearly better than another?
- Is the "jobs versus technology" argument raised by the manufacturing staff valid?
- Should people be concerned about losing their jobs with this option?
- What about the concerns over scheduling of particular runs? When would this most likely become an issue? How could this impact be measured or estimated?

EXERCISE 4: Spray Painting Analysis

Instructions for Exercise 4

This exercise builds on the experience of Exercise 3 by introducing more complicating factors. Again, use the process and cost information to assess the feasibility of the following pollution prevention project. Incorporate both quantitative and qualitative analyses. Use worksheets provided on the following pages. Answers can be found on pages 000 and 000. This exercise is complicated enough to warrant the use of a spreadsheet and the advanced TCA. If you decide to enter the data on a spreadsheet, you will have to include the equivalent annual annuity calculation. You may also wish to add additional cost categories or more detail to observe how the profitability of each option changes as the costs increase or decrease.

1. Use the provided role information to fill in the table. Calculate capital costs of each alternative, total operating costs, and incremental cash flow.
2. Calculate the present value (PV), the net present value (NPV), and the equal annual annuity (EAA) for each alternative.
3. Answer the following questions:
   a. At which points in the analysis does the attractiveness of the different options change?
   b. Which alternatives can be eliminated based on their net present value?
   c. What additional factors would affect the feasibility of these projects?

Magnatronix Corp.

Magnatronix Corp. produces spindle plates and shock casings from high-carbon steel for use in the aerospace industry. The company's pollution prevention team has decided to focus on opportunities in the spindle and casing painting operation. The paints currently used have a high percentage of methyl ethyl ketone (MEK) as a carrier base. Usage of these paints resulted in

114,352 lb of MEK stack emissions and 12,864 lb of MEK containing paint waste shipped off-site. Annual costs are $81,750 for paint purchases and $8,400 for disposal. After researching the available alternatives, vendors of alternative painting equipment were asked to make presentations to the pollution prevention team. The company's discount rate is 10%.

Option 1: The Bright Boy Alternative

Bright Boy Chemical Company proposed switching to a reformulated water-base coating using a newly developed carrier called oxy-2-butadiene, trade named MIRAGLO. This new chemical is not on EPA lists nor is it taxed under the Montreal Protocol for CFC Phaseouts. Bright Boy claims the chemical is self-cleaning, decreasing the amount of downtime for clearing clogged paint-delivery hoses. Paint equipment upgrades such as high-volume, low-pressure paint guns, a new water wall to catch paint overspray, new mixing equipment, and sparkless pumps, and fittings would make MIRAGLO an easy and painless switch from MEK. MIRAGLO requires no downtime for fine-tuning the system nor extra drying time. Furthermore, the system requires minimal retraining of workers and can use the existing ventilation system. The vendor says this will have a lifetime of at least 4 years.

The Bright Boy sales rep gave the team the MSDS for MIRAGLO and a list of companies currently using this product. Each team member has looked over the information and has some information on this option.

Purchasing

- System changes would cost $2,700 and $37,200 in equipment purchases.
- MIRAGLO paint is 40% cheaper than MEK ($49,050).
- Despite vendor's claims, all MEK must be removed from the system, the lines cleaned, and the waste MEK disposed. This results in an additional $5,000 initial cost.

Environmental Management

- MIRAGLO has a low volatile organic compounds (VOC) content, allowing it to be discharged to drain rather than drummed as a hazardous waste.
- Water use would increase to $1,500/yr, placing the facility in a significant discharger category.

- Waste MIRAGLO would need to be treated by the factory's pretreatment system, treatment chemicals cost $3,000/yr and testing $1,800/yr.
- There is evidence that oxy-2-butadiene causes mutations in laboratory animals.

Engineering

- Changeover would require a new water-discharge permit and a one-time permit fee of $500.
- Electricity use would drop to $3,520.

Manufacturing

- Production costs would increase by 5% due to smaller delivery lines and fittings. This would cost the company $15,000/yr in lost production time compared to current practices.
- Maintenance costs would be around $100/yr.

Option 2: The PC-Tech Alternative

The P2 team also interviewed Powder Coating Technologists, Inc. (PC-Tech), manufacturer of PCT, an electrostatic powder coating process. A sales representative for PC-Tech said the product meets all Mil-Spec coating requirements without risks to workers. PC-Tech manufacturers a whole line of chemicals and powder coating machines and will assist in custom designing a system. The sales rep also said PCT generates no hazardous waste. The vendor says this project easily could have a 7-year lifetime. Again, each team member had something to offer about this proposal:

Purchasing

- PCT chemical purchase costs would be $42,100/yr.
- The system would cost $150,000.

Environmental Management

- The oversprayed powders can be sifted and added back into the supply hoppers, eliminating hazardous waste disposal.
- Associated training costs are $20/yr.

Engineering

- The PCT system requires more floor space due to the conveyorized design, with ultrasonics and a hot-air dryer at the end, meaning $15,800 in changes.
- Production costs would increase by 7% because of the extra steps involved in preparing parts.

Manufacturing

- Maintenance costs would be around $200/yr.
- Electricity use would double to $8,000.

Option 3: The Rön-Jen Alternative

The final option evaluated by the Magnatronix team involved Rönhauser-Jennings, a manufacturer of new high-tech alloys and machining technologies. Rön-Jen demonstrated the use of a new metal alloy that could replace the high carbon steel. This alloy does not corrode and has a higher tensile strength than steel while being lighter per cubic foot.

If Magnatronix installed a computerized machining cell that utilizes lasers for cutting and polishing, the need to clean and paint parts would be eliminated. The new system could be designed to perform the lathe work, cutting, drilling, and welding operations. Additionally, the laser's precision finish was more durable than the paint. The vendors say the machine has at least a 10-year lifetime.

Purchasing

- Design and purchase of new machinery costs $228,000.
- The new alloy eliminates the painting process, saving $81,750 in chemical purchases, but it is 8% more expensive than high-carbon steel, resulting in $24,700/yr in materials costs (place in the chemical category).

Environmental Management

- No hazardous wastes are generated.
- Built-in safety lockouts assure worker health and safety.

Engineering

- The unit requires rearranging the floor layout and building a new room. Alterations would cost $152,000.
- Electricity costs would rise to $6,500/yr.

Manufacturing

- Eliminating the need to clean and paint parts would decrease production costs by 35%, saving $105,000/yr over current costs.
- Maintenance on new system would be $350/yr.
- Electricity usage would increase to $6,000/yr.

See Tables 8.13 to 8.15.

Answers for Exercise 4

See Tables 8.16 to 8.18.

Financial Feasibility

The three proposals have significantly different capital costs and different lifetimes. If the decision were made solely on purchase cost, MIRAGLO would win. Net present value cannot be used for comparison in this case because of the different lifetimes of each project.

This requires the use of the equal annual annuity method to accurately compare their profitability. EAA is calculated by dividing each project's net present value by the corresponding present value factor in Appendix A. For example, MIRAGLO's net present Value is $3,733; the present value factor for this 4-year project at 10% cost of capital is 3.1699. Dividing the NPV by this

■ TABLE 8.13 Capital Cost Worksheet

Capital Costs	MEK Spray	MIRAGLO Water	PCT Powder	Rön-Jen Alloy
Equipment purchase	N/A			
Disposal of old process	N/A		($5,000)	($5,000)
Research & design	N/A			
Initial permits	N/A			
Building/process changes	N/A			
Total capital costs	N/A			

■ TABLE 8.14 Operating Cost Worksheet

Operating Cash Flow		MEK Spray	MIRAGLO Water	PCT Powder	Rön-Jen Alloy
Chemical purchases					
Waste management	Chemicals				
	Testing	($1,000)			
	Disposal				
Safety training/equipment		($40)	($20)		
Insurance		($10,000)	($14,000)	($2,000)	0
Chemical use fees					
Filing paperwork time		($500)			
Annual permitting		($300)	($150)	0	0
Production costs	% inc./dec.				
	$/yr	($300,000)			
Maintenance	Time	($250)			
	Materials	($75)			
Utilities	Water	N/A			
	Electricity	($4,175)			
	Gas/steam				
Total annual operating cost factor					

■ TABLE 8.15 Cash Flow Summary Worksheet

Cash FLow Summary	MEK Spray	MIRAGLO Water	PCT Powder	Rön-Jen Alloy
Total operating costs				
Incremental cash flow				
−Depreciation				
Taxable income				
Income tax (40%)				
Net income				
+Depreciation				
After-tax cash flow				
Present value				
Total capital cost				
Net present value				
Equivalent annual annuity				
Infinite horizon NPV				

■ TABLE 8.16 Answer Key 1

Capital Costs	MEK Spray	MIRAGLO Water	PCT Powder	Rön-Jen Alloy
Equipment purchase	N/A	($37,200)	($150,000)	($228,000)
Disposal of old process	N/A	($5,000)	($5,000)	($5,000)
Research & design	N/A	0	0	0
Initial permits	N/A	($500)	0	0
Building/process changes	N/A	($2,700)	($15,800)	($152,000)
Total capital costs	N/A	($45,400)	($170,800)	($385,000)

■ TABLE 8.17 Answer Key 2

Operating Cash Flow		MEK Spray	MIRAGLO Water	PCT Powder	Rön-Jen Alloy
Chemical purchases		($81,750)	($49,050)	($42,100)	($24,700)
Waste management	Chemicals	0	($3,000)	0	0
	Testing	($1,000)	($1,800)	0	0
	Disposal	($8,400)	0	0	0
Safety training/equipment		($40)	($20)	($20)	($20)
Insurance		($10,000)	($14,000)	($2,000)	0
Chemical use fees		0	0	0	0
Filing paperwork time		($500)	0	0	0
Annual permitting		($300)	($150)	0	0
Production costs	% inc./dec.	0	5%↑	7%↑	35%↓
	$/yr	($300,000)	($315,000)	($321,000)	($195,000)
Maintenance	Time	($250)	($100)	($200)	($350)
	Materials	($75)	0	0	0
Utilities	Water	N/A	($1,500)	0	0
	Electricity	($4,175)	($3,520)	($8,000)	($6,500)
	Gas/steam	N/A	0	0	0
Total annual operating cost factor		($406,490)	($388,140)	($373,320)	($226,570)

factor yields an equal annual annuity of $1,177/yr for the life of the project. If this project were replaced every 4 years with an identical system, forever, the Infinite Horizon NPV would be $1,177 divided by the cost of capital (10%). Based on the EAA method, the Rön-Jen alternative looks most profitable.

■ **TABLE 8.18 Answer Key 3**

Cash Flow Summary	MEK Spray	MIRAGLO Water	PCT Powder	Rön-Jen Alloy
Total operating costs	($406,490)	($388,140)	($373,320)	($226,570)
Incremental cash flow	N/A	$18,350	$33,170	$179,920
−Depreciation	N/A	($11,350)	($24,400)	($38,500)
Taxable income	N/A	$7,000	$8,770	$141,420
Income tax (40%)	N/A	($2,800)	($3,508)	($55,568)
Net income	N/A	$4,200	$5,262	$84,852
+Depreciation	N/A	$11,350	$24,400	$38,500
After-tax cash flow	N/A	$15,550	$29,662	$123,352
Present value	N/A	$49,133	$144,406	$757,948
Total capital cost	N/A	($45,400)	($170,800)	($385,000)
Net present value	N/A	$3,733	($26,393)	$372,948
Equal annual annuity	N/A	$1,177	($5,421)	$60,695
Infinite horizon NPV	N/A	$11,770	($54,210)	$606,950

Teacher's Notes for Exercise 4

If you choose to walk through the exercise, note the following points:

- Which alternative looks the most promising based solely on equipment purchase costs? (One might choose this option if no other analysis is done—MIRAGLO.)
- Which alternative looks the best after the annual savings have been calculated? (Again one might conclude that this is the right choice if no further analysis is done—MIRAGLO).
- Which are still feasible projects after the net present value calculations? (Now how do we choose between the two alternatives—MIRAGLO and Rön-Jen.)
- Which one is the best considering the information provided? (Rön-Jen. What considerations might change this? Layoffs of workers? Ability to raise capital?)
- Notice the inconsistent electrical usage cost in Option 3. Sometimes your data will conflict and a judgment must be made. In this case, the higher cost was chosen to provide a smaller cost savings and a more conservative financial analysis.
- Notice the $5,000 to clean out the lines and get rid of the extra MEK, even with MIRAGLO's drop-in replacement. Oftimes, these

preparatory steps are overlooked in the initial setup phase. Replacement chemicals usually require a fresh start with no traces of the previous chemicals.

- MIRAGLO causes mutations in rats, a possible indicator for mutagenicity in humans. What are the trade-offs of using a new chemical?
- Notice the lack of information in some instances. How can this be corrected? Is this a problem? Will it significantly impact the analysis?

CHAPTER 9

TCA AND CONTINGENT LIABILITIES

This chapter could save your company a lot of money and possibly your job. The decision makers of your company rely on you to maintain environmental compliance and minimize the losses from environmental liabilities. These liabilities represent one of the largest threats to a corporation. Contamination of a community drinking water supply, illegal dumping hazardous wastes, or evacuation of neighborhoods due to accidental releases of gases can result in multimillion dollar fines, personal injury claims, and even criminal prosecution. To avoid these scenarios, you need to know

- what a contingent liability is and if you can control, reduce, or eliminate it
- how to evaluate, rank, and estimate the costs of these liabilities
- how to include contingent liabilities costs in a TCA

This chapter will explain all these points and provide tools to help you quantify these liabilities. In case you are among those who think, "Oh, it will never happen at my facility," the following examples provide sobering evidence that these events are not limited to large corporations such as the *Exxon Valdez* oil spill or Union Carbide's Bhopal gas leak.

Forklift Accident

A forklift operator hit a hydrochloric acid above-ground storage tank, releasing dangerous fumes into the neighboring community. Area residents and businesses were evacuated and several people were treated for fume inhalation at a local hospital. Bodily injury and business interruption claims exceeded $460,000.[1]

Cause: Operator error

Preventable: Yes

How: Better employee training

Improper Labeling

A manufacturer of dry cleaning products stored incorrectly labeled drums of raw materials outside the company's building. Neighbors called the fire department after noticing a thick, whitish-yellow cloud emanating from the drums. The firefighters read the labels on the drums and began spraying water on them. The drums exploded and the chemicals reacted to form a cloud of sulfur dioxide. Forty plaintiffs filed three lawsuits to recover over $3 million in damages for injuries suffered from exposure to the sulfur dioxide cloud.[2]

Cause: Insufficient internal environmental compliance

Preventable: Yes

How: Improve internal environmental auditing

No Containment

A waste-storage area stacked with drums of a caustic substance did not have secondary containment. The caustic substance eroded the drums and spilled onto the ground and into an adjacent creek. Remediation of the site involved removal of the contaminated soils from the premises and dilution of the waste in the creek. Total cleanup costs exceeded $100,000.[3]

[1]Examples modified from Environmental Commercial Services, Inc., liability case studies, March 14, 1997. Web site is http://www.ecsuw.com.

[2]Ibid.

[3]Ibid.

Causes: Lack of preventative maintenance and internal environmental compliance

Preventable: Yes

How: Increase maintenance and environmental compliance activities

Improper Spill Response

Periodic chemical spills from mixing and blending areas of a warehouse were washed down the drains and entered the septic system. Investigation of drinking well contamination at neighboring residences indicated the warehouse's septic tank as the originating source. Soil and groundwater cleanup costs exceeded $500,000.[4]

Causes: Insufficient worker training and poor management attitude

Preventable: Yes

How: Better training and show management the costs of these liabilities

HOW MUCH DO I NEED TO KNOW?

The subject of environmental liability could be easily a book in itself. The volume of information and its technical complexity may explain why many people fail to take a serious look at contingent liabilities. This chapter provides only the essential information to help you begin evaluating your plant's environmental liabilities. To start the evaluation and then select projects to eliminate contingent liabilities, you need to know the following:

- the basic vocabulary of the environmental liability industry
- the options available for qualitatively or quantitatively evaluating liabilities
- the steps necessary to include contingent cost information in a TCA

WHAT ARE CONTINGENT LIABILITIES?

A liability is an obligation or responsibility resulting from some action. It usually entails the payment of money to agencies, contractors, plaintiffs, or civil or criminal prosecution. Contingent liabilities are often referred to as "intangible liabilities" because the obligations attached to the liability are

[4]Ibid.

unknown in scope and timing. The author prefers the word "contingent" because these liabilities have not yet occurred and are contingent on certain activities happening (i.e., the valve will not turn off, the spill will reach the storm drain, etc.). These sources of risk can be defined and steps taken to reduce or eliminate them. Thus, the liability becomes more defined and less "intangible."

Two broad categories of contingent liabilities exist: historical and preventable. *Historical liabilities* are all existing liabilities incurred from *past practices* of the manufacturing facility. It is irrelevant whether events have occurred that force a payment of the obligation; it only matters that the conditions for these obligations have been created already. These liabilities will include all the hazardous wastes ever sent to a landfill and any chemicals and materials that have been spilled on the ground or released to the air or water. All of these "pollution exposures" (the term given to environmental contamination events by the insurance industry) will have to be rectified eventually, whether it is the closing and monitoring of a landfill, pumping and treating a contaminated aquifer, or settling personal injury claims of workers or community residents.

From a prevention-oriented perspective, little can be done to avoid these "pollution exposures." Once the activities that create the liability have occurred, the liability exists.

The second category are liabilities that the company has the potential to create in the future. These *preventable liabilities* accrue due to the *future practices* of the company, such as continuing to ship hazardous waste to a landfill, any future releases of emissions to the environment, exposure of workers to toxic chemicals, and other activities. Because the actual "pollution exposure" has not occurred, it can be minimized or eliminated given appropriate action.

Preventing these pollution exposures minimizes the creation of new and or additional contingent liabilities and their associated contingent costs. Costs from preventable contingent liabilities are more readily estimated, and, more importantly, the cost differences between current practice and a pollution prevention project can be shown. This book focuses on this second category (preventable liabilities), because TCA can best show the strong correlation between pollution prevention and reducing these types of liability.

Types of Contingent Costs

Contingent costs are the costs that a company incurs to meet the obligations of the contingent liabilities. These costs would be incurred only if an event triggers the obligation of the contingent liability. The following lists a few contingent costs categories.

Penalties and Fines A company can incur a penalty or fine through a variety of infractions, some as minor as not having adequate aisle space between hazardous-waste drums or as serious as dumping cyanide out behind the building. Generally, penalties and fines will be incurred if the company refuses to fix the identified problem, ignores the warnings of the agency, or is found at fault due to negligence or willful intent. Fines can be calculated using complex matrices that weigh the relative risk of the action or can be stipulated in a federal or state regulation as a certain amount per day of violation. For example, noncompliance with the discharge limits of the National Pollution Discharge Elimination System permit can carry a fine of $25,000/day for each day that each limit has not been met.

Legal Expenses and Superfund Lawsuits Millions of dollars can be spent defending your position in a Superfund case, and in some cases, it would have been less expensive to simply clean up the contaminated site. This is not to say that a company should never defend itself against false accusations, but sometimes the best way to win a conflict is to not be involved in the first place. No matter who ends up paying for the cleanup, there will be significant costs incurred by all parties involved.

Remediation Superfund litigation aside, the labor, treatment, and disposal of contaminated materials can run into six, seven, and eight digits. Most remediation occurring today is the result of negligence and ignorance of the past 100 years. Today, storage, treatment, and disposal regulations force a more conscientious approach. Future remediation will focus more on plant decommissioning and cleanup of accidental spills and releases.

Workers Compensation and Personal Injury Claims On-the-job accidents and injuries from handling wastes are sources for these costs from within the company. The mere presence of dangerous materials, such as reactives, corrosives, and compressed gases, creates the environment for an accident to occur. Pollution prevention projects really shine when the reduced accident rate can be attributed to the elimination of a particular chemical or process.[5]

[5]Most companies currently have a "________ days without accidents" program. The more progressive of these programs provide incentives for the workers to take care and prevent accidents, thus reducing the number of on-the-job accidents and consequent personal injury claims. In one facility, the savings in insurance premiums resulting from the decreased accident rates is given to the employees at the end of each year in a lottery. Only those employees who have not had any accidents can participate.

Contamination outside the facility may result in community or citizen suits for personal injury. The public often sides with the individual not the corporation, so the outcomes of these law suits typically favor the plaintiff, which can impact the image costs of the corporation.

Natural Resource Damage Contamination of watersheds and natural habitats creates deficits that are particularly difficult to quantify. These models are tailored to the complexity of a particular type of event. Models exist that help place a dollar value on damage caused by oil and chemical spills, groundwater contamination, gas leaks, explosions and fires, and soil contamination of heavy metals and solvent volatiles. In order to evaluate your company's contingent costs, several models may have to be employed.

Chapter 12 discusses several efforts underway to measure the more permanent macroeconomic impacts of industry and societal activities on the ecosystem. The continuing impact to Prince William Sound in Alaska from the *Exxon Valdez* oil spill is just one example. Hundreds of millions of dollars and volunteer time were spent to clean up the damage, yet scientists say some areas will be damaged for the next decade or longer.

Image Costs These costs are perhaps one step more elusive than contingent costs. For example, how can one estimate the impact of a press release from an environmental group that is flashed around the world on the Internet? What is the cost of losing favor with your host community? These types of costs can be broken down into two basic categories, corporate image and community relationship.

CORPORATE IMAGE This would be the overall perception of the company within the business community and the country at large. It is one of the few intangibles over which a company has any kind of control. Larger corporations have the luxury of being able to spend large sums of money polishing their corporate image through aggressive public relations campaigns and advertising.

Smaller companies must rely on a "no news is good news" type of image. That is, since the company is not featured on the front page of a newspaper for polluting, the image is intact.

As an indirect measure of image costs, the budgets of these campaigns could be included in the TCAs for eliminating polluting processes or chemicals. Thus the process that can damage the image is charged for the cost of maintaining or improving the image.

For example, McDonalds fast food corporation underestimated the effectiveness of the consumer boycott on Styrofoam food containers. After several years of persistent effort, activists finally achieved their goal and McDonalds agreed to change to paper-based wrappers. The pressure of the boycott had cost the corporation enough poor publicity to make it worth the change.

COMMUNITY RELATIONSHIPS The image cost in this case is the value of the perceptions of customers, investors, suppliers, employees, host communities, and others. Smaller businesses may be interested in the trend of their larger brethren to select only the cleanest, most above-board companies as suppliers. Corporations such as AT&T, IBM, and the big three auto manufacturers have rigorous guidelines their suppliers must follow to retain the larger company's business. In addition to product quality guidelines, some of these buyers are stipulating compliance with applicable environmental regulations. A supplier in litigation with a state agency or the federal government now stands to lose revenues from sales as well as paying the legal expenses for environmental noncompliance. The increase in ISO certified companies brings additional pressure to bear on environmental image. Lack or loss of certification can mean lost opportunities to do business overseas.

The most straightforward method of evaluating image costs is to qualitatively rate the impacts of various negative scenarios (i.e., groundwater contamination of a local aquifer, chemical explosion, clear-cutting of old growth forests, etc.). Although not financially quantified, this rating can be used as additional information when a project's TCA is reviewed. You can use the identical qualitative methods outlined later in this chapter for contingent costs to evaluate the image costs as well. An example of image cost evaluation is provided in that section.

HOW ARE CONTINGENT COSTS RELEVANT TO TCA?

TCA provides environmental cost information as a decision criteria. Contingent costs are another level of environmental costs and can be easily incorporated in the TCA. The contingent costs are a third tier of costs (conventional and hidden costs are the first two tiers) that can be added into the TCA when needed. Usually, the contingent costs result from an existing process and can be reduced or eliminated through the implementation of a pollution prevention alternative.

For example, vapor degreasers using chlorinated solvents represent a potential liability because the chlorinated solvent may be spilled on the ground or into the sewer, causing environmental harm. If discovered by state or federal regulators, this accident would result in penalties for breaking the law, costs for lawyers, bad publicity locally or nationally, and costs for contractors to come and clean up the spill and remediate the soil or water.

Every year the company chooses to use this chemical rather than switch to a less toxic substance, it increases the probability that an accident will occur. The probability increases because the complacency of a routine leads people to forget about the risks of the materials they work with and the number of potential opportunities for a spill increases. For example, there would be potential for a spill for the following:

- unloading the solvent at the loading dock
- transferring the solvent to the storage area
- bringing the solvent out of the storage area and to the degreaser
- filling the degreaser with the solvent
- if the degreaser leaks while being used
- cleaning out the degreaser
- transporting the waste solvent from the degreaser to the waste-storage area
- loading the waste solvent onto the waste hauler's truck
- the truck is traveling to its destination

Using the solvent raises *nine* separate opportunities for a spill to occur. This does not include opportunities for worker health exposure risks or accidents that may impact stored chemicals, such as forklifts puncturing drums. The point is, if a suitable replacement for the chlorinated solvent can be found, the facility could avoid the liabilities of using solvents.

HOW HAVE CONTINGENT COSTS BEEN DEALT WITH HISTORICALLY?

By and large, there have been three dominant responses to the subject of contignent costs. These responses are not limited to the business sector, but are also seen in the environmental enforcement agencies, technical assistance programs, and lending institutions.

1. *"Hide Under the Blanket."* The first response to contingent costs is the "hide under the blanket" routine, denying that any potential liabilities

exist. For any number of reasons (complacency, acceptance of risk, ignorance, or outright fear), people using this response choose to ignore the potential costs and take a wait-and-see attitude.

2. *"Yell 'Fire' in the Movie Theater."* The second response seeks to motivate companies to eliminate their preventable liabilities by waving the red flag of contingent costs, yet not offering help to determine what these liabilities are. Whatever the reason for not expanding on the subject (labor shortfalls, perceived complications of the task), this action comes across as a scare tactic, and weakens the argument for eliminating liabilities.
3. *"Keep It at Arms Length at All Times."* Rather than being in denial or afraid to address the costs, this response stresses uninvolvement and distance. These people may be so sensitized to liability issues they will leave the room before you finish saying the word. Past involvement with litigation or bankruptcy can be the root of this skittish behavior.

Why Try to Evaluate Contingent Costs?

These attitudes about contingent costs cover the spectrum from denial to outright paranoia yet somehow miss the mark on actually knowing what the contingent costs might be. Without knowing what the costs might be, how can one adequately prepare for them or select alternatives to eliminate them? Evaluating contingent costs demystifies them, allowing the costs to be seen for what they are rather than as specters looming on the horizon.

FEDERAL REGULATIONS REGARDING LIABILITIES

Publicly owned company's need to consider the legal implications before attempting to evaluate liabilities. This section discusses the finer points of how investors are protected from risks and how to determine your company's responsibilities to disclose these risks to investors. Whether your company is publicly or privately held, evaluating contingent liabilities requires a firm understanding of all legal requirements and restrictions. This book does not outline what each company should or should not do, it only presents the methods and allows you to determine their suitability.

Who Governs Investor Risk?

The Securities and Exchange Commission (SEC) is charged with overseeing the trading and investing practices of corporate America. The SEC promulgates regulations regarding the size and type of transactions that can occur, who may or may not be involved, and requires all *publicly held companies* to annually report their financial activities in a report known as the 10K. Regulation S-K specifies the contents of the 10K report and item 101(c)1(xii) describes the responsibilities of publicly traded corporations to disclose:

> the material effects that compliance with Federal, State and local provisions which have been enacted or adopted regulating the discharge of materials into the environment, or otherwise relating to the protection of the environment, may have upon the capital expenditures, earnings, and competitive position of the registrant and its subsidiaries.[6]

What Do The Regulations Require?

The SEC regulation by itself is not completely clear and therefore was clarified in a 1979 Interpretive Release. The following three major points are paraphrased from the release. You are encouraged to review the exact language of this release or have an attorney interpret it for you. The interpretive release states:

1. All environmental compliance capital expenditures for periods beyond 2 years shall be disclosed if the company developed or received estimates for them. These would include the cost of achieving compliance with future regulations, any penalties or fines anticipated, and the assumptions and methods used in reaching the estimates.[7]
2. The company must report any administrative proceedings pending or contemplated by government agencies and the amount of relief sought by those actions. The company is also required to report any "cease and desist" orders delivered by these agencies.[8]
3. Other information that may be of interest to investors, such as company environmental policies, planned projects, and so forth, should be included in the 10K.[9]

[6]Securities and Exchange Regulation S-K, Item 101(c)(1)(xii), Federal Securities Laws, Regulation Subsection 229.101, Paragraph 13,011, Commerce Clearing House, Inc., 1985.

[7]1979 SEC Lexis 621, Securities and Exchange Commission, September 27, 1979, p. 5.

[8]Ibid., p. 6.

[9]Ibid., p. 7.

How Do These Regulations Impact TCA?

The S-K disclosure regulations are not clear enough regarding the preventable liabilities that are evaluated by TCA. However, with some discrimination, it is possible to plot a course of action, meet all legal requirements, and still include an evaluation of contingent costs within a TCA.

1. The first step is to determine how these regulations apply to your company. Consult your company's corporate attorney or chief financial officer to determine the ownership status of the corporation. Even if your company is a privately held corporation, state regulations may govern your corporation. Check with the corporate attorney or the secretary of state to determine requirements under state statutes.
2. Once your company's obligation to disclose contingent costs has been determined, you can choose to either qualitatively or quantitatively evaluate them. Depending on the choice, you may be required to report any contingent costs that are probable, likely to occur, and have been estimated. The decision to quantify the contingent costs must be made at the highest level of the corporation and with its full knowledge and support. This book cannot tell you which method is appropriate for your company, only what methods are available.

METHODS FOR EVALUATING LIABILITIES

In order to eliminate or reduce the preventable liabilities of your firm, you must have methods for evaluating and prioritizing the contingent costs. Each liability represents risk and uncertainty, either to the current operating practice or the investment being considered. The tools available to help you make decisions amidst risk and uncertainty include qualitative analysis, which only describes the relative preference for one event to occur over another, and quantitative analysis, which involves probability and modeling event outcomes. Depending on your desire to forecast financial values and how well defined the liability is, one approach may be more appropriate than another.

Using Qualitative Methods to Evaluate Contingent Liabilities

The qualitative approach to evaluating and prioritizing risks and liabilities fits well with companies that may have issues with SEC requirements for disclosure of all liabilities. These methods will assist in the decision-making process by ranking risks in a subjective fashion.

■ **FIGURE 9.1 Pareto Chart for Part Defects**

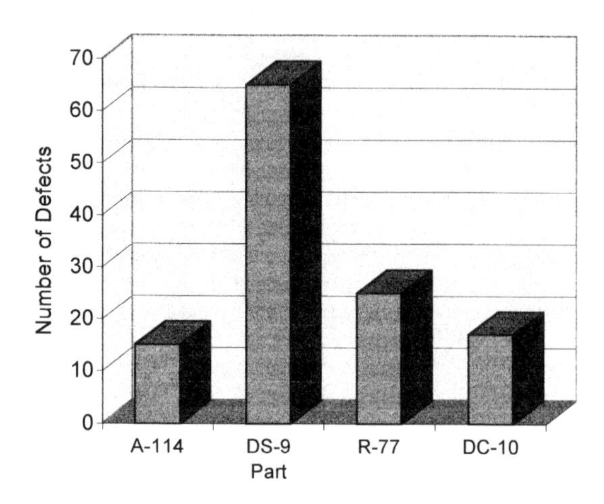

Pareto Analysis of Risk and Probability One of the old standbys of the quality field is the Pareto diagram. From this diagram has sprung the expression "80% of your problems are caused by 20% of your process," also known as the 80/20 rule. Pareto analysis is typically used in analyzing process defects and visually showing the number of each type of defect. The quality team members then can see which defects occur most often. Pareto charts can identify the most important issues, help select a starting point for process changes, and identify underlaying causes of problems.[10] Figure 9.1 illustrates a typical Pareto chart. The chart shows the relative number of defects for particular products. Clearly, the DS-9 part number has a much higher number of defects than all others, prompting the inquiry: Why? Perhaps the answer does not actually lead to an improvement in the production process, but it provides a starting point for other questions and analysis. For example, it could be that the defect rates on the other parts are improperly reported.

Pareto charts can assist in qualifying the various risks and seriousness of each environmental issue. To use the chart, one must first gather data on the types of risk or contingent costs relevant to the process being analyzed. This data could represent the number and type of operator accidents, the number of spills during maintenance functions on the process, comparative data on waste or emissions generation from each process in the facility, and any other function or process-related data that would show aspects of risk due to the types of chemicals or materials used, created, or disposed.

[10]Peter Capezio, *Taking the Mystery out of TQM.* Hawtherne: Career Press, 199X p. 173.

■ FIGURE 9.2 Expected Changes in Operator Injuries After the P2 Project

	A-line	DS-line	R-line	DC-line
Current	15	65	25	17
New Proc.	12	45	25	12

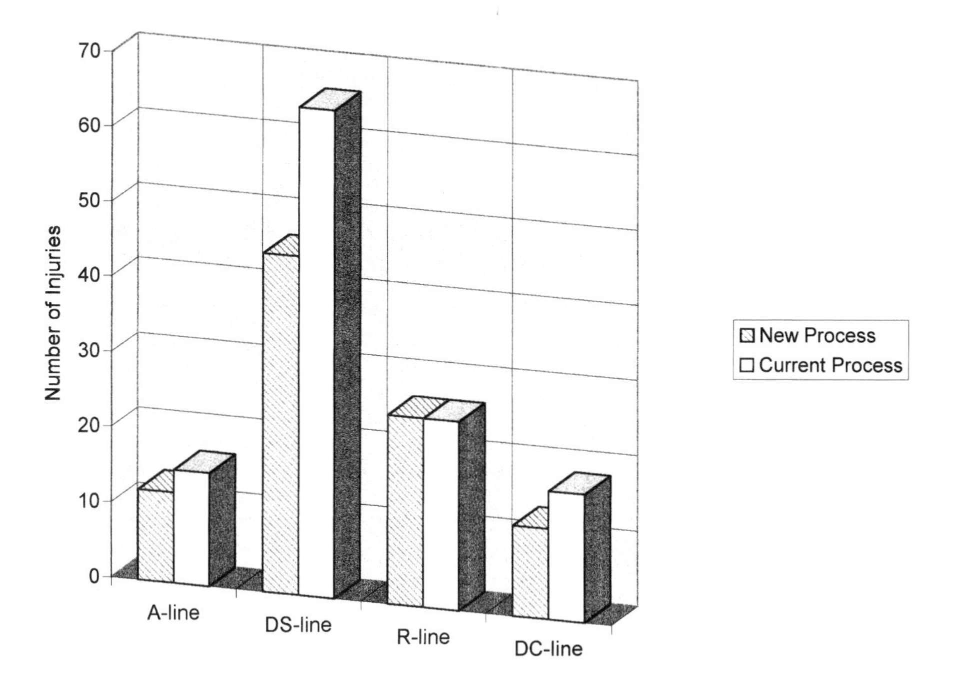

The next step is to structure the Pareto diagram to show how the alternative processes or process changes will compare to the existing process in terms of identical type of data. Most spreadsheet packages can create 3-D bar graphs several layers deep. A variation of the diagram shown in Figure 9.2 might be useful for showing expected reductions in worker accidents for a new process. The graphical portrayal of the data clearly shows the reduction in injuries on the DS-line. Of course, the diagram alone does not explain how the reductions will occur, but it does convey a powerful image of the relative reductions expected.

Risk Matrix The final tool for qualitative analysis of contingent costs compares potential risks through a matrix format. This format is often used after brainstorming sessions to select a final alternative as measured against specific criteria. In many ways, it is a less complex model of the analytical

■ TABLE 9.1 Contingent Cost Trade-Off Matrix

	Alternatives				
Risk	**Outsource**	**Automate**	**Substitute**	**Improve Rinse**	**Weight**
Chemical spills	N	Y	N	Y	2
Worker injury	N	N	N	Y	1
Treatment chemicals	N	Y	Y	N	2
Landfill sludge	Y	Y	Y	N	3
Release to sewer	N	Y	Y	N	1
	3	8	6	3	

hierarchy process. The matrix approach allows the display of a large array of criteria and alternatives. Criteria can be weighted to reflect their relative importance and then each alternative evaluated as to whether it meets the criteria.[11] Weighting the criteria helps break ties in the event that alternatives meet the same desirable criteria. Table 9.1 illustrates the basic principle of creating a matrix for the contingent costs of a small plating line.

Risks Associated with Plating	***Contingent Costs from the Risk***	***Relative Weight***
Chemical spills	Cleanup costs	2
Worker injury	Workers compensation claims	1
Treatment chemicals	Cleanup costs	2
Landfill sludge	Remediation/closure costs	3
Release to sewer	Litigation costs	1

Alternative A: Outsource the plating

Alternative B: Automate the plating operation

Alternative C: Substitute less hazardous chemicals

Alternative D: Improve the rinsing process and reduce dragout

See Table 9.1 for the matrix.

In this example, the alternative with the lowest score is the most desirable. Note the decision switch was a simple "yes" or "no." Here a "no" means the alternative would not create the same level of risk as the current practice.

[11] G. Thuesen and W. Fabricky, *Engineering Economy*, 6th ed. Englewood Cliffs, NJ: *Prentice* Hall, 1984, p. 491.

Some judgment must be used to select one of the two choices. Look at "Landfill Sludge" under "Substitute," for example. Here the assumption was made that a substitute product would still create sludge that would require landfilling and thus continue to generate some contingent costs. Under the fourth option, this same criterion was given an "N" because the concentration of contaminants was expected to be below the regulatory limits that would declare the sludge hazardous and thus decrease the contingent costs. A more accurate variation of this matrix might include a relative weight for how much each alternative will change the expected contingent costs. For example, the fourth alternative in the example might eliminate landfill sludge concerns and also reduce some risk of spills. Perhaps this risk reduction is half of the current level of risk, and therefore a 0.5 cofactor is justified.

Analytical Hierarchy Process The Analytical Hierarchy Process (AHP) was developed over 20 years ago by Thomas Saaty as a complex decision-making tool that allows integration of rational and intuitive information. AHP is a predictive decision-making aid where the user assigns priorities to specified objectives. Pairwise comparisons among the objectives allow the user to confer preferences for one objective over another. Manipulation of the ranked pairwise comparisons generates a matrix that is solved to find eigenvalues (the basic central tendency of the value weight or fundamental frequency of the information). One benefit of this process is the ability to include inconsistencies in judgment, such as ambiguity, risk, conflicting interests, and combinations of qualitative and quantitative information.[12]

AHP can be used in conjunction with TCA to provide a more balanced view of the alternatives being considered. For example, when addressing several pollution prevention projects simultaneously, AHP can be used to select the option that best meets the company's goals. These goals may be to select the option that most lowers liabilities, improve image, or other subjective trade-offs. By coupling this process with a TCA, you can achieve a comfortable estimate of the financial and intangible impacts of the decision.

The following example illustrates the application of AHP to the selection of one of three options for improving a company's corporate environmental image. The three options follow.

[12]Analytical Hierarchy Process history (from company portfolio), Pittsburgh: Expert Choice, Inc., 1997.

1. Donate money to the local environmental activist group.

Pros	***Cons***
Free publicity from news	Short-term memory of group and public
Tax-deductible contribution	Loss of capital in fiscal year

2. Go to zero discharge on all the processes within the facility.

Pros	***Cons***
Reduces operating costs	Capital cost is very high

3. Run an extensive advertising campaign and make no other tangible changes.

Pros	***Cons***
Longer run time than donation	May be perceived as "green washing"

Each option has distinct trade-offs that are well into the intangible realm. Some might consider donating money to an environmental group to be "warm and fuzzy," whereas other people might consider it a bribe. Does the free publicity from donating money provide a higher-quality image boost than regular advertising?

Suppose there is no way to quantify the costs involved with going to zero discharge before the commitment has to be made. Also suppose it would be a voluntary project, the timeline is undetermined, and it allows the company to spread its investment over several years and achieve publicity at each milestone.

The third option provides the highest public visibility, but also could be perceived as "green washing," that is, intentionally trying to make a bad performance record look good by putting some spin on the public relations. What are the trade-offs here if the advertising campaign backfires on the company?

The AHP process allows you to weigh one possible outcome against another and assign values to each. For example, which is more important, to have high-quality publicity or to have a long-running advertising campaign? Or how much more important is it to reduce waste generation compared to minimizing capital expenditures? Figure 9.3 is a computer screen shot from an AHP software package called ExpertChoice.[13]

[13]This example was developed and analyzed using Expert Choice Software Version 9.0. Pittsburgh: Expert Choice, Inc., 1997.

■ FIGURE 9.3 Analytical Hierarchy Process Example

To improve coporate environmental image

Compare the relative PREFERENCE with respect to: GOAL

	MAXTIME	MINCOST
PUBQUAL	3.0	2.0
MAXTIME		1.0

Row element is __ times more than column element unless enclosed in ( )

Abbreviation	Definition
Goal	To improve corporate environmental image
PUBQUAL	Attain high quality publicity
MAXTIME	Maximize length of time image remains positive
MINCOST	Minimize the amount of capital spent

PUBQUAL	0.550	
MAXTIME	0.210	
MINCOST	0.240	

Inconsistency Ratio = 0.02

(*Source*: Easy Choice, Inc., 1997)

This example used the previously stated goal (improve corporate environmental image) and the following three objectives;

1. Attain high-quality publicity, that is, significant detail about the positive accomplishments of the company and present a credible message to the public.
2. Maximize the length of time this image remains in people's minds.
3. Minimize the amount of capital spent.

Each of these objectives grew out of the "pros" and "cons" listed under each possible project, some of which were outlined earlier. Each pro and con becomes a subobjective and all components are compared to each other through the use of graphical, verbal, or mathematical evaluation methods. Algorithms distribute a weight to each subobjective and create a score for the alternatives. These values are then graphed to show the strengths of each project with respect to each objective. In this example, the zero-discharge project has the highest quality of publicity that also can be spread over the longest time. This project fails in realizing the goal of minimizing cost, but overall, scores better than the other two projects.

An analysis like this has the flexibility to adapt to the changing conditions in and around the facility. For instance, the grass roots environmental group may decide to kick off an antitoxics program just as the company is deciding to donate money to them. Is a donation to the group still appropriate, and how do the potential repercussions weigh against the other options? All of these variables can be handled through the AHP process.

The preceding example shows one application of AHP for intangible image costs. AHP also could be applied directly to pollution prevention projects like the following. A company desires to determine which alternative paint formulation would provide the most benefits. Concerns and objectives include:

- Quantity of paint required, coverage, cost
- Disposal issues
- Retraining workers
- Potential unknown risks of new chemicals
- Relative ease for cleanup of painting equipment
- Trade-offs of decreases in toxic fumes for carrier evaporation rates

Each of these has to be "decomposed" into a set of objectives and subobjectives and weighted against one another. The software package could provide a graph showing the weighted ranking as additional information for the TCA.

Limits to AHP and Other Qualitative Methods

The lack of numerical values is the biggest limit of all the qualitative contingent cost evaluation methods. Not having a number to plug into the TCA leaves the question "How much will it cost me?" unanswered. These costs are potentially more significant than all other cost categories, and, as such, carry weight in the decision-making process. It is important to pair the qualitative analyses of contingent costs with a TCA of the conventional and hidden costs to cover all the relevant cost categories.

QUANTITATIVE APPROACHES TO COMPUTING CONTINGENT COSTS

There may be times when you desire to quantify the amount of contingent costs for particular processes. You then could show in dollars that an alternative will help eliminate or avoid the need to pay out the contingent costs at some point in the future. There are many pieces to the puzzle of contingent cost quantification. Here are some of the cost categories that need to be considered when evaluating contingent costs:

- Soil and waste removal and treatment
- Groundwater removal and treatment
- Surface sealing
- Personal injury
- Economic loss
- Real property damage
- Natural resource damage
- Litigation costs

Using Preexisting Models to Compute Contingent Costs

Over the years, various federal agencies and private companies developed computer models that simulate spills and calculate the cost of cleanup. Many of these models are available for free or a small fee and can provide a first approximation for the costs involved with a chemical or oil spill. You can also add in fines and penalties for these events by using the methods described later in this chapter. Finally, you could couple all of this with a probability model such as Monte Carlo analysis and develop a range of costs, a mean, and a variance. This number would tell you on average what it will cost for a pollution exposure event, and when it might occur.

The following models calculate contingent costs based on an assumed or already occurring pollution exposure event. They make assumptions regarding the size of the spill and the type of damage that would occur by using average values derived from governmental studies of cleanup costs. The models cover events such as the following:

- Groundwater contamination on-site or at the receiving TSDF
- Natural resource damage from oil spills in coastal waterways
- Settlement of Superfund liability issues by paying your company's "fair share".

These models are used by the agencies that developed them, so your computed costs will be close to what a prosecuting agency would seek from your company in the event the spill or release actually occurs.

Estimating Corrective Action Costs Based on Waste Toxicity and TSDF Location Suppose you want to show the advantages of a new process change that can eliminate or drastically reduce the amount of hazardous waste sent off-site to the a hazardous-waste landfill. You know at some point the landfill will be full and must be closed; any groundwater contamination coming from the landfill will need to be remediated. Because your company shares in the liability of the landfill, it will be contributing some percentage toward the corrective action costs. What factors contribute to the costs of remediating this Treatment Storage and Disposal Facility (TSDF)?

- Payment of claims for natural resource damage
- Payment of claims for third-party lawsuits and bodily or property damage
- Federal agency cost recovery in administrative orders and negotiated settlements under the Comprehensive Emergency Response Compensation and Liability Act (CERCLA)
- Remediation costs at the facility site under the Resource Conservation and Recovery Act (RCRA); these include surface sealing, fluid removal, and treatment
- Remediation costs for third-party property for which the waste generator is liable under CERCLA

From reports such as the EPA's *Estimating Costs for Economic Benefits of RCRA Compliance* and DOD's *A Comparison of the True Costs of Landfill Disposal and Incineration of DOD Hazardous Wastes*, you can determine a unit cost for remediation of a generic TSDF site.

The TSDF to which your company currently sends its wastes can be rated against this average. If it failed to have the average containment and maintenance measures, then the unit cost for remediation could be adjusted upwards.

You would have to determine the amount of future waste to be sent to the landfill under the current process. This amount, represented as a fraction of the total amount of waste stored at the TSDF, can provide a cost-sharing percentage for the site's remediation. The distance of the TSDF from a public water supply, the types of materials you sent there, and their rate of migration through water all could be used to determine a time period in which the public water supply would be contaminated.

These numbers would be combined to produce a total contingent cost for the amount of waste expected to be sent to the landfill. This cost is then added to other TCA costs.

This was the approach General Electric used to develop its "Financial Analysis of Waste Management Alternatives Model." In 1987, GE teamed with ICF Incorporated to develop their own methods to evaluate the profitability of waste minimization projects and their impact on future liability, specifically, costs related to TSDF liability from groundwater contamination.[14]

This software is designed for waste minimization calculations, including choosing which TSDF will result in the lowest contingent costs. The spreadsheet and manual are available from General Electric for a mod est fee.

Contingent Costs for Oil Spills of Coastal Waterways Perhaps you are more interested in projects that reduce your facility's liability regarding oil and chemical spills to coastal waterways. A pollution prevention project that creates safer transfer procedures or containment technologies, or eliminates the need for fuel oil, may reduce the probability of a spill and the associated contingent costs. In this case, you may want to explore the software developed by the Department of the Interior.

The Department of the Interior has developed a user-friendly computer software package that models natural resource damage from spills of hazardous substances or oil. This program is used by federal, state, and Native American tribal natural resource trustees to calculate monetary damages for spills in coastal and marine environments. The Natural Resource Damage Assessment Models for Coastal and Marine Environments (NRDAM/CME) and for Great Lakes Environments (NRDAM/GLE) provide the user with a presumption of correctness in a court of law, requiring the accused party to disprove the damage assessment.

Users supply data such as wind conditions and the amount and duration of the spill. Four submodels calculate the following:

- How the spill moves and distributes on the water surface, along shorelines, in the water column, and in sediments
- Direct mortality and direct loss of production resulting from short-term exposure to the spill
- Indirect mortality and indirect loss of production resulting from food web losses
- Whether various restoration actions are warranted (such as dredging and refilling of sediments, or washing of sand and gravel)
- The cost of these restoration actions

[14]Monica Becker and Allan White, *Total Cost Assessment: An Overview of Concepts and Methods*. Boston: Northeast Waste Management Officials Association, July 1991, p. 7.

- The value of lost use of injured resources for activities such as hunting, fishing, birdwatching, and going to the beach[15]

A Natural Resource Damage Assessment Example On October 14, 1995, a Connecticut company accidentally discharged 5500 gall of No. 4 heating oil into a storm drain that emptied into the Quinnipiac River. A transfer pump seal broke over the weekend, allowing oil from the storage tank to leak into a floor drain that emptied to the river. Although the company had been told by the state DEP to fix the drain, no interim measures were taken while plans for changes to the piping were being reviewed by the DEP. The storm drain discharges to a section of the river flanked by tidal marshland that provides habitats for many species of waterfowl, migratory birds, and fish.

The spill was contained after special booms were brought in overnight by air from California. During the next 15 days, over 3,500 gall of oil were recovered. The DEP canceled the early waterfowl hunting season in the marsh area, a popular location visited by more than a 100 hunters per day. All the oil-soaked marsh grass in the effected area had to be mowed to encourage regrowth. Oil-soaked wildlife had to be cleaned up and returned to undamaged marsh areas. There were also concerns that oyster beds at the mouth of the Quinnipiac River would be damaged, resulting in economic losses for fishermen.

The company cooperated with the state DEP and the Department of the Interior during the cleanup and agreed to pay all associated costs. The state attorney general's office levied a fine of $51,000 to compensate for the loss of wildlife and recreational and economic use of the area.[16]

In the preceding example, the NRDAM model calculated sums for the cost of restoration and loss of hunting, fishing, and use of the resource. By constructing a hypothetical future spill event with this model, you can create a future economic damage cost. This sum represents a contingent cost for a spill at some point in the future. Using Microsoft Excel and a probability modeling add-in, you could find a mean and variance on the likely year for a spill to occur. The contingent cost would be presumed to occur in that year and could then be discounted back to the present day. This sum could be incorporated into the TCA as a contingent cost for the preventable liability of any future use of oils or hazardous chemicals.

[15] The Department of Interior's NRDA model for Coastal Marine Environments or for the Great Lakes Environment can be downloaded from the Midwest Marine Science Board web site http://www.msc.nbs.gov/nrda/nrdahtml.

[16] Connecticut Department of Environmental Protection, Press Releases dated October 16 through November 24, 1995, CTDEP, Hartford.

The cost estimates of these models have been controversial due to size of the economic damage figure calculated. This works to your favor because a larger cost estimate will only serve as a stronger motivator. Remember also that this estimate will be placed somewhere out in the future and discounted by a set interest rate.

Settlement of Superfund Liability Claims Based on "Fair Share" Analysis The third modeling method calculates a minimum allowable amount used to settle a facility's involvement in Superfund litigation. The U.S. EPA developed CASHOUT[17] to calculate a lump-sum share of cleanup costs for a potentially responsible party (PRP), allowing them to cashout of the settlement process. This program estimates future costs for cleanup and maintenance of the site and can be adapted to hypothetical situations, allowing you to estimate contingent costs for such a scenario. One of its benefits is the ability to set the cleanup project start date and the date of the first payment by the PRP. The calculated value represents the total remediation and maintenance cost of the site, only a portion of which belongs to the PRP. By approximating the ratio of waste the PRP sent to the site versus the total amount of waste in the site, a portion of this cost can be allocated to the PRP. Figure 9.4 shows a typical calculation from the CASHOUT program.

Limits to Computer Models Computer models are based on assumptions that may not apply to your situation. Models, especially those developed by government agencies, are always criticized for shortcomings. The most common complaint is that the cost estimates are far too high. Computer models designed by others may not be flexible enough for your needs, or cover every cost category desired. Always examine the assumptions made by the model and include them in your TCA summary.

Using Research-Based Averages to Compute Contingent Costs

If your primary concern is compliance costs within your own facility or image costs resulting from pollution exposures, you may be able to use cost figures developed from extensive surveys of environmental managers around the country. If you are familiar with statistical sampling, you can estimate

[17]CASHOUT can be obtained through the National Technical Information Service, U.S. Department of Commerce, 5285 Port Royal Road, Springfield, VA 22161, 703-487-4650, or on the Internet at http://es.inel.gov/oeca/models/CASHOUT.html.

■ FIGURE 9.4 Demonstration of CASHOUT Software

```
CASHOUT, VERSION 1.5

ARCHIE, MONTANA                                         APRIL 2, 1997

PRESENT VALUE OF CLEANUP COSTS AS OF COST COMMENCEMENT DATE:

RECURRING EXPENDITURES EVERY  10 YEARS         $      16307
ONE-TIME EXPENDITURE                           $   10500000
ANNUAL EXPENSE FOR  30 YEARS                   $      30387
                                                  =========
TOTAL AS OF JAN, 1998                          $   10546693


PRESENT VALUE OF CLEANUP COSTS AS OF PRP PAYMENT DATE:

RECURRING EXPENDITURES EVERY  10 YEARS         $      15835
ONE-TIME EXPENDITURE                           $   10196359
ANNUAL EXPENSE FOR  30 YEARS                   $      29508
                                                  =========
TOTAL AS OF JUN, 1997                          $   10241702


->->->->->->THE SUPERFUND CASHOUT CALCULATION ABOVE <-<-<-<-<-<-
                 USED THE FOLLOWING VARIABLES:

USER SPECIFIED VALUES
---------------------
1. SUPERFUND SITE NAME =       ARCHIE, MONTANA

2. RECURRING EXPENDITURES EVERY 10  YEARS     = $       4000  1998
DOLLARS
3. ONE-TIME EXPENDITURE                       = $   10500000  1998
DOLLARS
4. ANNUAL EXPENSE FOR  30 YEARS               = $       1500  1998
DOLLARS
5. PRP PAYMENT DATE                           =            5, 1997
6. SUPERFUND CLEANUP COSTS COMMENCEMENT DATE  =            1, 1998

STANDARD VALUES
---------------
7. ANNUAL INFLATION RATE                      =        1.60%
8. DISCOUNT RATE                              =        4.50%
```

(*Source*: U.S. EPA, 1997)

how accurate these figures might be for your facility, based on the size of your facility and the complexity of its manufacturing processes.

In 1990, the U.S. Environmental Protection Agency published the *Pollution Prevention Benefits Manual* through a combined effort of the Office of Policy Planning and Evaluation and ICF Incorporated.[18] This manual contains information on quantifying all levels of costs including the contingent costs

[18]The U.S. EPA *Pollution Prevention Benefits Manual*, Document EPA/230/R-89/100, is available from the National Technical Information Service, U.S. Department of Commerce, 5285 Port Royal Road, Springfield, VA 22161, 703-487-4650.

and less tangible costs such as corporate image and consumer relations. The manual provides formulas and default values for many of the cost categories used in TCAs. Several of the contingent cost tables have been reproduced in Appendix B for convenience. An example follows.

Personal Injury

Formula:

$$\text{Personal injury} = a\ (\text{average claim}) \times b\ (\text{potentially affected population})$$

Variables:

a = average claim per person for lost time due to disability, mortality, medical costs related to illness, and medical monitoring costs
= \$56,000 per person (default value)

b = potentially affected population (number of people)
= 15,000 (worst-case default value)
= 3,500 (typical-case default value)
= 10 (best-case default value)

To use this information in the TCA, you will need to know the surroundings of the site that will experience the chemical leak or spill, whether it is your facility, or the TSDF that handles your wastes. Population density, proximity of drinking-water supply, and sensitive ecological areas all contribute to the total cost for fixing pollution exposures. All of the default values in the manual are based on research conducted in 1988 and 1989. If you are uncomfortable using data that is not current, contact and survey the environmental lawyers and consultants in your area to gather new local averages. If another division of your company is currently involved with remediation or litigation over a pollution exposure, examine their cost data and compare them to your hypothetical situation. Remember, the goal is to develop a cost component of the TCA for a hypothetical event in the future.

Using Civil Penalty Matrices to Compute Contingent Costs

For specific contingent costs such as a single spill, contamination event, or chronic noncompliance with regulations, the civil penalty arising from this situation can be estimated and used as a contingent cost. This cost may be large enough even to prove the worth of the pollution prevention project without further analysis.

Companies on the receiving end of the penalty wonder how the agency could possibly come up with such a tremendously large figure. Typically, the penalty is composed of two parts, a "gravity-based" penalty to reflect the seriousness of the noncompliance, and an "economic" penalty to eliminate the financial benefits of not complying.

This second portion of the penalty actually helps justify including penalty calculations in a TCA. If the regulating agency is going to punish your facility by taking away profit gained by not complying, it follows that this "profit" should be added into the benefits side of any project that could eliminate the penalty-causing liability.

How Penalties Are Calculated Most agencies have a penalty matrix that helps enforcement officers develop each component of the penalty. See Table 9.2. The gravity-based component starts as a subjective rating of the risks to people, environment, surface or groundwater quality, and the amount and nature of the wastes. This is coupled with a subjective classification of the degree of deviation from the regulation. The enforcement official would gauge based on past compliance history and whether the event was a minor, significant, or substantial deviation.

If you desire to include the potential penalties resulting from a future event, check with your state EPA to see if the penalty matrix is public information. You can obtain the penalty matrices from your regional EPA office through a Freedom of Information request.

To illustrate how to use such a matrix, a state penalty assessment methodology will be applied to an example event at a large manufacturer of cast metal products.

A large manufacturer die-casts alloy forms and passivates the parts in a chromating process. Spent chromating solutions exceed the toxicity characteristic leachate procedure (TCLP) guidelines for chromium, requiring the solutions to be manifested off-site as hazardous waste. On the third shift, a drum of this solution is punctured on the loading dock by the company's forklift operator because it was not properly stored. The drum spills out onto the pavement and down into the storm drain. Fearing a reprimand, the operator washes the spill down the drain and tells no one. Several days later, the the state Department of Environmental Conservation receives a call from a hunter who spots dead ducks floating in the estuary. Wildlife biologists determined the ducks died from heavy-metal poisoning and the source is quickly traced to the company. An enforcement action is started and a penalty is calculated as follows:

TABLE 9.2 Generic Penalty Matrix Compliance With State Hazardous-Waste Regulations

Category	Score			
	0 = N/A or Nil Risk	1 = Low Risk	2 = Moderate Risk	3 = High risk
Population	No significant pathway to the public	Low impact on public; unlikely pathway to public; distant receptor	Moderate actual or potential impact on - public; short pathway to nearby receptor	Actually or potentially very dangerous; high chance of exposure; nearby receptor
Environment	No significant pathway to environment	Low impact on environment; unlikely pathway to environment; distant receptor	Moderate actual or potential impact on environment; short pathway	Severe or widespread impact; nearby delicate ecological situation
Amount of waste	Insignificant	Several pounds or gallons	Tens of pounds or gallons	Hundreds or thousands of pounds or gallons
Nature of waste	Harmless	No lasting effect; biodegradable; only minor irritant on exposure	Toxic, explosive, reactive or corrosive; injurious on exposure	SARA extremely hazardous substance or other actually toxic material
Groundwater or surface water quality	Did not or cannot reach groundwater or surface waters	General industrial area; water quality class GB, GC, D, C, SB, or SD	Generally clean area; water quality class GA, A, or SA	Actual or potential prime drinking-water supply; water quality class GAA or AA

■ TABLE 9.3 Degree of Deviation

Minor	Most parts of the regulation are met, but some parts are not fully complied with.
Moderate	Significant deviation from parts of the requirement, but meeting other parts as the regulation intended.
Major	Substantial deviation from the requirement, ignoring a requirement, or making protection inoperative.

The gravity-based component is derived from the matrix of Table 9.2, where there is no significant pathway to the public (0 points), but "high risk" or severe impact to delicate ecological situation (3 points). At least 10 gall was spilled (2 points) and the substance was considered toxic on exposure (2 points). Finally, there was impact to surface-water-quality class SC.

Next, the degree of deviation (see Table 9.3) is determined to be "Major" because the operator did not follow the spill protocol outlined in the facility's contingency plan. The potential for harm in this case is considered to be "Moderate" because the total score of all risk categories is 7. Combining this score with the "Major" degree of deviation yields a gravity-based penalty of $10,625. See Table 9.4.

The economic-based component of the penalty consists of the avoided and the delayed costs of not complying. Complying in this case is construed to mean storing the waste in a proper location with secondary containment, having a spill-response kit ready, and training workers to respond properly. All of these costs must be based on the current costs at the time the noncompliance was discovered. The avoided costs are those that never can be incurred because the company cannot go back into time and perform the tasks. For this example, the avoided costs are as follows:

Item	*Cost*
Labor to move drums as they are filled	$200
Not conducting weekly inspections	$1,400
2 yr of annual training for workers	$1,200
Total	$2,800

Next, the total is adjusted by the corporate tax rate (assume 40%) and the interest charged by the IRS on delinquent accounts, using the equation

$$\text{Adjusted avoided cost} = \text{avoided cost} \times (1 - \text{effective tax rate}) \times (1 + I)^n$$

■ TABLE 9.4 Scoring Matrix for Penalties

Potential for Harm	Degree of Deviation		
	Major	Moderate	Minor
Major score, 11–15	$25,000	$20,000	$15,000
	$22,500	$17,500	$13,750
	$20,000	$15,000	$12,500
Moderate score, 6–10	$12,500	$8,750	$5,000
	$10,625	$6,875	$4,000
	$8,750	$5,000	$3,000
Minor score, 1–5	$3,000	$1,750	$500
	$2,375	$1,125	$300
	$1,750	$500	$100

where I is the IRS interest rate (assume 12%), and n is the number of years the facility has been out of compliance. For this example, n is 2 yr, yielding

Adjusted avoided cost component	$2,107

Delayed costs are items that must be purchased now to attain compliance. Because these costs were not incurred when they should have been, the agency takes away any interest earned on the capital not spent. The delayed costs are

Purchasing secondary containment devices	$1,000
Purchasing a spill-response kit	$1,500
Training for workers this year	$1,200
Delayed cost component	$3,700

$$\text{Interest on delayed costs} = [\text{delayed costs} \times (1 + I)^n] - \text{delayed costs}$$

Interest on delayed costs	$941

The total penalty for this event is, therefore,

Gravity component	$10,625
Avoided costs	$ 2,107
Delayed costs	$ 4,641
Total penalty	$17,373

Once the penalty is calculated, it can be added in totality or in part to the TCA figures. Or at your discretion, it can be included in the summary section along with the assumptions. Many times this is as effective as adding it to the calculation because the perceived impact of a penalty is often greater than its monetary value.

Calculating the Economic Benefit Portion Using a Computer Model The U.S. EPA has developed a computer software program called BEN, which calculates the economic portion of the penalty.[19] This portion is the profit gained by the company during the period they were out of compliance with the regulation. It uses an NPV formula to arrive at a current dollar penalty amount. The user can enter a variety of variables and set the dates of noncompliance and payment. Figure 9.5 shows a typical calculation run by the BEN model.

The sample calculation in the figure represents a hypothetical penalty for not having conducted RCRA refresher training for employees at the fictitious Genconn facility. The costs for this future noncompliance event total $3,275 in 1997 dollars. This figure could be added to any TCA that might eliminate or reduce the need to conduct RCRA refresher training. The assumptions of the scenario are as follows:

- At some point in the future, the facility will fall behind in its RCRA training requirements (this is one of the more common RCRA violations).
- An environmental agency will discover this and fine the company for failing to maintain compliance.
- The company will pay the fine in the year 2001, realizing the contingent cost of $4,883 (2001 dollars).

Although you may not agree with the assumptions or calculations behind either BEN or Cashout, the results provide significant numbers through the same process that the state and federal agencies use. At the minimum, using the same models that compliance enforcement officials use allows the company to anticipate the size of the penalty when the violation is discovered. For TCA, these models make computation easier and more in line with the

[19]BEN can be obtained through the National Technical Information Service, U.S. Department of Commerce, 5285 Port Royal Road, Springfield, VA 22161, 703-487-4650, or on the Internet at http://es.inel.gov/oeca/models/BEN.html.

■ FIGURE 9.5 Demonstration of BEN Software

```
GENCONN                                    BEN Version 4.3 APRIL 3, 1997

A.   VALUE OF EMPLOYING POLLUTION CONTROL ON-TIME AND OPERATING IT FOR
ONE USEFUL LIFE IN 1997 DOLLARS                                    $8598

B.   VALUE OF EMPLOYING POLLUTION CONTROL ON-TIME AND OPERATING IT FOR
ONE USEFUL LIFE PLUS ALL FUTURE REPLACEMENT CYCLES IN 1997 DOLLAR$11480

C.   VALUE OF DELAYING EMPLOYMENT OF POLLUTION CONTROL EQUIPMENT BY 48
MONTHS PLUS ALL FUTURE REPLACEMENT CYCLES IN 1997 DOLLARS          $8205

D.   ECONOMIC BENEFIT OF A  48 MONTH DELAY IN 1997 DOLLARS
 (EQUALS B MINUS C)                                                $3275

E.  THE ECONOMIC BENEFIT AS OF THE PENALTY PAYMENT DATE,   48 MONTHS
AFTER NONCOMPLIANCE                                                $4883

     ->->->->->->THE ECONOMIC BENEFIT CALCULATION ABOVE <-<-<-<-<-<-
                  USED THE FOLLOWING VARIABLES:

USER SPECIFIED VALUES
     1A. CASE NAME = GENCONN
     1B. PROFIT STATUS = FOR-PROFIT
     1C. FILING STATUS = C-CORPORATION
     2. INITIAL CAPITAL INVESTMENT (ONE TIME) = $2000  2001 DOLLARS
     3. ONE-TIME NONDEPRECIABLE EXPENDITURE =  $0
4. ANNUAL EXPENSE = $1500  2001 DOLLARS
5. FIRST MONTH OF NONCOMPLIANCE = 1, 1997
6. COMPLIANCE DATE =1, 2001
7. PENALTY PAYMENT DATE =1, 2001

STANDARD VALUES
     8. USEFUL LIFE OF POLLUTION CONTROL EQUIPMENT = 15 YEARS
     9. MARGINAL INCOME TAX RATE FOR 1986 AND BEFORE = 49.6 %
     10. MARGINAL INCOME TAX RATE FOR 1987 TO 1992 = 38.6 %
     11. MARGINAL INCOME TAX RATE FOR 1993 AND BEYOND = 39.4 %
     12. ANNUAL INFLATION RATE = 1.6 %
     13. DISCOUNT RATE: WEIGHTED-AVERAGE COST OF CAPITAL = 10.5 %
```

(*Source*: U.S. EPA., 1997).

cost that might be accrued in the event of noncompliance. Both BEN and CASHOUT are available for free from the U.S. EPA.

Limits to Using Penalty and Fine Calculations Calculating expected penalties does not cover all the costs associated with the act of noncompliance. Most significantly, the labor of the company's employees in attaining compliance is left out. In some situations (such as training all company employees), this cost can be substantial. Penalties are objectively determined based on an outside review of the facility's compliance history and the nature and scope of the event. As an employee of the corporation, you may not share the same perspective and could underestimate the penalties. Try to be as objective as possible when calculating penalties for your company, and consider adding in the labor required by your company to attain compliance.

Liability Insurance Premiums as a Surrogate for Contingent Costs

The environmental insurance industry is a rapidly growing field providing loss coverage for environmental professionals, construction contractors, and manufacturing facilities. Prior to 1970, commercial liability insurance covered all acts within the facility, and as a result implied coverage of gradual and sudden pollution exposure. Because of catastrophic losses connected with this implied coverage, insurance companies instituted the "Pollution Exclusion" after 1973 and an "Absolute Pollution Exclusion" on policies written after 1986.

Manufacturing companies were left without coverage during a time when Superfund litigation and clean-up were beginning to drain many corporate bank accounts. In response to this need, several companies have developed "Pollution Legal Liability" insurance policies that cover sudden and gradual pollution exposure remediation costs including defense costs. The insured (the manufacturer) agrees to pay a set policy fee and retain a deductible for a set liability limit. If the amount of coverage needed is too large for the primary insurer, a reinsurer may be brought in to transfer the additional liability.

An example of this coverage is a $40 million liability limit for a $10,000 minimum annual premium with a $25,000 self-insured retention (deductible). To establish this policy, the manufacturer must be assessed by the insurance company to discover all environmental risks. The insurance company may require the insured to perform risk-reduction activities. The policy then becomes an annual operating cost for the company.

In this manner, the contingent costs of all the firm's environmental liabilities are converted to this annual operating cost. This annual operating cost can be included when conducting a TCA for projects that would reduce or eliminate the liabilities covered under the insurance policy. Insurance providers periodically reevaluate facilities to adjust for changes in risk levels. Most of the time, the risk levels increase, but occasionally the premiums are lowered if the insured has significantly reduced risk to the facility.

One example of this is a large printing company that uses nitroglycerin as a component in the inks. The company consumed enough ink to warrant a tank car of nitroglycerin being parked on the railroad siding next to the factory. The risks from the chemical in this tank car are fairly obvious and influenced the premium paid by this company. The company then switched over to an aqueous printing system and eliminated the tank car. The insurance provider reduced the company's premium when the tank car was eliminated.

Limits to Using Insurance Premiums Many factors go into premium rate setting. The insurance coverage for liabilities might be just one part of a much larger policy. Changes to production processes might not eliminate the risks that account for the largest portions of the premium. With the exception of the environmental liability insurance companies, there are no known resources that collect all the data on compliance infractions, spill and release occurrences, and costs associated with them. Interview the policy underwriters or agent and determine the risks that impact the premium the most. This can both focus the pollution prevention efforts of the company and indicate when it is most advantageous to include insurance premiums as a measure of contingent costs.

Evaluating contingent costs can be time-consuming, but counting the contingent cost reductions will accelerate project payback and increase the IRR. The most time-efficient approach would include contingent costs as a last resort after gradually including more costs into the TCA until a profitable state has been reached.

CHAPTER 10

INTEGRATING TCA WITH OTHER MACROMANAGEMENT TOOLS

The majority of this book deals with Total Cost Assessment (TCA) as an isolated tool for financial decision making. No environmental manager ever operates in isolation but must continually adjust to the latest management philosophies, whether it is ISO 14000, Total Quality Management, Activity Based Costing, Theory of Constraints, or Life Cycle Analysis. Although constant adjustment and continual change can be inconvenient and frustrating, the most pragmatic view is to see these recent developments of management techniques as having many parallels to pollution prevention and TCA.

Total Cost Assessment helps you make decisions between dissimilar projects based on their estimated savings. These projects often originate through some type of periodic review of the company's production process, such as a pollution prevention opportunity assessment, ISO 14000 review, or part of a continuous improvement process like Total Quality Management. TCA can be used as an accounting counterpart to these continuous improvement processes to help identify areas within the company that consume significant portions of operating capital.

Rather than outrightly rejecting either the TCA approach or any of the other tools, try to look at how they overlap or complement each other. The flexibility of the TCA method can be adapted to any of these other tools to complement their decision-making or process evaluation phases. In fact, TCA addresses some issues that these other tools leave out. Certain philosophies focus more on production processes and less on environmental management, providing opportunity for TCA to fill this void. In short, there is room within the manufacturing facility for all approaches to work. This chapter highlights how TCA fits with some of the more popular management philosophies and where they might conflict.

THEORY OF CONSTRAINTS

This section answers the following questions:

- What is the Theory of Constraints?
- How is it compatible with TCA?
- How is it in conflict with TCA?

What Is the Theory of Constraints?

The Theory of Constraints (TOC) was developed over 17 years ago by Eliyahu Goldratt, a management consultant trained as a physicist. By approaching industrial production problems in the same fashion as scientific problems, Mr. Goldratt adapted scientific methods to business management. Essentially, the theory states: The flow through the entire production process is limited by the maximum output of the slowest process or step of the system. By optimizing this constraint, the whole system becomes more efficient, and more able to turn inventory into throughput.

Specifically, the Theory of Constraints is a practical description of the Socratic method, or the processes of thinking. According to the originator of the theory, by knowing how to think, we can better understand the world around us; by better understanding, we can improve.

Central to the concept of TOC is the acknowledgment of cause and effect. The processes of TOC provide steps combining cause and effect with experience and intuition to gain knowledge. Three basic components make up the TOC:

1. Problem-solving tools to logically and systematically answer the three essential questions of ongoing improvement: "What to change?" "To what to change to?" "How to cause the change?"
2. Daily management tools to improve management skills, such as communication, effecting change, team building, and empowerment

3. Application-specific processes, such as for production, distribution, marketing and sales, and project management, among others [1]

Implementation of the Theory of Constraints at any facility requires full participation from all employees and managers. This process may be used to review production processes and identify bottlenecks in production, and adjust work flow to optimize throughput capacity at the bottlenecks.

Although focused primarily on production and work flow, the Theory of Constraints is instructive to the pollution prevention opportunity assessment process. The theory asks several questions (stated earlier), the answers to which will be found on the shop floor.

- What to change?
- What to change to?
- How to effect the change?

Through these questions, a process for improving the production flow develops. Ultimately, any improvement should, according to Goldratt, "simultaneously increase throughput, and decrease inventory and operational expense"[2] (*Note:* Each of these terms has a specific definition, somewhat different than traditional management definitions.) A similar set of questions and processes could be used to find pollution prevention opportunities and analyze them using TCA.

For example, if a company desired to increase production capacity without exceeding its environmental permit limits (air pollution or water-discharge limits), it might use the TOC method to define the limiting step (the regulation or permit) and what to change (specific products or inputs) that create more capacity at that limiting step (perhaps lowering output of pollution through substitution of a material or chemical input).

How Is TCA Compatible with the Theory of Constraints?

In a broad sense, both methodologies are attempting to change a corporate structure that lacks specific elements for improved performance. For example, in the Theory of Constraints, the process of identifying the constraints and changing corporate measures of success provide a clearer picture of how the company operates. Under TCA methodology, the analysis gives a clearer picture of where the money is being spent on environmental compliance and

[1] *From The Goldratt Institute, New Haven,* via the Internet, 1996.

[2] Eliyahu and Goldratt, *The Goal.* Great Barrington, MA: North River Press, 1992, p. 67.

where the savings are. Both tools help managers make more informed decisions. Both tools therefore can work synergistically to promote change in corporate culture and shop practices.

The Theory of Constraints is actually more related to pollution prevention than Total Cost Assessment. Pollution prevention is an environmental management theory that takes a large-scale, whole-facility view as does the Theory of Constraints. Other macromanagement theories such as Design for the Environment, Sustainable Manufacturing, and Industrial Ecology all attempt to encircle the factory and minimize environmental impact. Unfortunately, most of them do not propose a people-based solution to achieve the changes. The instructions are often written as if there will be no resistance to the change. The Theory of Constraints offers tools to overcome the common barriers to change. For this reason, TOC could be very successful in the environmental arena as a management tool.

If followed to the letter, the Theory of Constraints would require a change in the way inventory and expenses are tracked. This realignment of accounting categories has some similarities with the total cost approach of pulling some categories out of overhead and making the analysis specific to a particular process.

Where Do TCA and the Theory of Constraints Conflict?

The Theory of Constraints takes a macroview of the company in that it is more important that the whole facility be making money than every process, or every worker be making money (i.e., being productive all the time). This aspect runs parallel to the pollution prevention belief of a multimedia approach to environmental management, however, it does not directly relate to TCA. TCA is connected to the profitability of the facility only so far as a sound business decision made using the TCA method may help keep a company in the black.

There may be some conflict between the two approaches if TCA were applied as the accounting rule in the facility and the Theory of Constraints was the operating rule in process optimization. This conflict would arise because the Theory of Constraints methodology rejects all accounting and financial number "games" except for three basic indicators. These indicators are as follows:

- *Throughput:* The rate at which a company generates money through sales

- *Inventory:* All the money the company has invested in things that it intends to sell
- *Operational Expense:* All the money the company spends to turn inventory into throughput[3]

According to the theory, these indicators can replace all others, and to generate these indicators requires some regrouping of traditional accounting cost categories. For example, inventory warehouse space would be considered part of operational expense and not part of an asset that might be sold. Or in another instance, the value added to a product from labor is not treated as part of inventory, but instead counts as an operational expense. The theory does not explore where the costs for environmental compliance fit within these three indicators. They might be placed as operational expenses, because they are a necessary part of staying in business. In this case, the expression "staying in business" refers to the process of turning inventory into throughput. These few sentences hardly can do justice to the scope of TOC theory. TOC's subtle differences in operational definitions and ways of viewing production result in staggering improvements.

A second potential conflict is indirectly raised by Caspari and colleagues[4] in the expression of concern over the Theory of Constraint's allocation of investment expense as a one-time charge in operational expenses. This allocation causes a large dip in the net profit reported for that period, not reflecting the savings or improvements achieved through TOC application. Caspari and colleagues advocate spreading the investment expense across the number of periods required to pay back the investment, rather than one period, and rather than depreciating over the entire lifetime of the purchased good. This payback allocation shows the purchase and payback of the investment over time and then the increased profit following the payback period.

The assertion that a nontime-value-of-money method be used to report on the current net profit and projected net profit creates a potential conflict between TOC and TCA in the accounting arena. Although commonly used as a secondary performance measurement tool, payback analysis has certain flaws that make it blind to cash flows. This relatively small conflict can be avoided by the careful application of payback analysis only during reviews of existing accomplishments. Decision making on future projects should still

[3]Ibid., p. 60.

[4]JohnA. Caspari, "Allocation of Investment Cost to Operational Expense: A TOC Approach," http://users.aol.com/caspari0/toc/RC0.HTM; 1994.

incorporate the net present value or another discounted cash flow method. If you are determined to use payback analysis, consider at least using one or two additional decision tools to round out the financial numbers.

ACTIVITY BASED COSTING

This section answers the following questions:

- What is Activity Based Costing?
- How is it compatible with TCA?
- Where does it conflict with TCA?

What Is Activity Based Costing?

Ever since the first accountants began keeping track of cash flows, there have been accounting ledgers. These ledgers classify costs into categories for the purpose of determining the profit made by the company. These categories do not relate very well to the activities that generate the revenue that creates the profits. For example, one expense category might be labor. This single line item encompasses all employees whether they sweep the floor, run a milling machine, or assemble parts. Even if the labor category is subdivided into assembly labor, milling labor, and maintenance labor, the tasks performed (such as deburring, tapping, painting, etc.) are excluded. Consequently, there is a data gap between the types of activities performed in manufacturing a product and the cost incurred in performing it. Without this linkage, there must be "leaps of faith" regarding how much it really costs to produce a product.

The upsurge of management theories advocating the empowerment of employees, teamwork, and redesign of everything under the sun creates a new demand on these old accounting systems. The "people" cry out for data that relate to their specific process, product, or activity. In many cases, the accountants have been unable to give it to them, not because they did not want to, simply because the accounting system was never designed to function as a two-way information process. That is, the small details of production go up to the front office, but only large companywide figures, if any, come back down.

Activity Based Costing (ABC) was developed in response to this shortfall and helps operations do its job by making cost information available at a level everyone can use in day-to-day decision making.

Activity Based Costing assigns costs to the activities within the company that result in product. For example, marketing, purchasing the materials, designing and drafting plans, milling, and painting are all activities

■ **TABLE 10.1 Comparison of Environmental Cost Categories to Environmental Activities**

Environmental Cost Categories		Comparable Environmental Activities	
Permits	Environmental compliance	Permitting	Reporting
Environmental compliance	Wastewater	Manifesting	Parts rinsing
Hazardous waste	Spent cleaners	Treating waste	Parts cleaning
Maintenance labor & parts	Personal protective equipment	Maintaining processes	Limiting worker exposure risks
Environmental compliance	Training	Inspecting	Training
Disposal costs	Taxes & fees	Disposing of wastes	Using chemicals
Utilities	Scrubber operation	Using energy and water	Treating waste

associated with creating a product. They all require resources and take time. These inputs can be measured and an average developed. This sum of all the average costs of these activities represents the cost to produce the product.[5]

How Is TCA Compatible with Activity Based Costing?

Both ABC and TCA seek to clarify accounting information, thereby helping all within an organization make better decisions. Those using TCA within a traditional accounting system have been, to some degree, trying to fit a round peg in a triangular hole. TCA is actually more akin to Activity Based Costing than to a traditional accounting system because many of the TCA costs are labor- or activity-related. With minor modifications, it is possible to modify TCA cost categories to reflect *activities* that generate pollution or impact the process being analyzed. By restructuring the TCA tables this way, a person in an ABC-run company could use the TCA methods for pollution prevention projects. Table 10.1 shows how environmental costs might be changed to environmental activities under ABC.

[5]From the Internet: http://info.rutgers.edu/Accounting/raw/ima/imabc.htm.

How might one construct a new type of ABC-based TCA form? (Excuse the acronyms.) Start by reviewing the basic concepts of Activity Based Costing. Costs are usually allocated to activities in four groups;

1. Unit Level: Activities performed on individual units
2. Batch Level: Activities that process batches of units
3. Product Sustaining: Activities that maintain the producing capacity for a product line
4. Facility Sustaining: Activities that maintain the functioning of the facility

Next, allocate all environmental activities to one of the four levels of activities. Then develop the costs of these activities as individual line items pulled from overhead. Examine each activity's "cost driver" to determine which products hold responsibility for the costs. Finally, assign the products a cost based on the number of events that occur for each activity assigned to each product the company manufactures. In most cases, there are products for which no environmental costs are incurred and other products that carry a heavy "environmental burden." ABC makes a point of not segregating environmental activities from production activities, as illustrated in the following example.

Suppose a company has two product lines, M5 and M1A. The activities required to produce each of these products are different for each and some of these activities are environmental management activities.

Activity	***Cost Driver***
Unit Level	
Degreasing	Only M5 product
Energy	Machine hours used
Disposal of hazardous waste	Only M5 product
Batch Level	
Spray painting	Only M1A product
Cleaning paint guns	Only M1A product
Air permitting	Only M1A product
Product Level	
Research and design	Number of subcomponents
On-site waste treatment	Both M5 and M1A
RCRA biennial reports	Only M5 product

Facility Level

Plant maintenance	% value added
Heating and lighting	% value added
EMAS	% value added [6]

This theory of cost allocation has been advocated or mandated by many of the state pollution prevention/waste minimization/toxics-use-reduction planning and reporting laws. These laws often require that the costs for using listed hazardous materials be allocated to the products that use them and that a "pounds of chemical per unit of production" be developed. Essentially, this shows that products using hazardous chemicals have additional costs attached to their manufacture that are usually placed in overhead and forgotten.

Activity Based Costing does exactly the same thing, but it only extends the benefit to all costs involved with producing products whether or not they are environmental costs. In the preceding example, the M5 and M1A products would have different total costs once all the costs were allocated. It appears from the list of activities and cost drivers that M5 uses more hazardous chemicals, creates a hazardous waste, and requires more environmental reporting. You could expect the unit cost of M5 to be higher than M1A if all other costs were equal.

Just as TCA helps clarify the costs of a current operation and a new alternative that may have additional environmental benefits, ABC shows areas within product manufacture that are not competitive. Perhaps the M5 product will be two or three times as costly to produce as M1A, solely as a result of environmental management issues. If the pricing structure for this product does not reflect this difference, then the company is subsidizing the production of M5 by shifting costs to the M1A product line. This then means that the M1A pricing structure is artificially high!

AT&T is one company that has embraced Activity Based Costing in environmental management. AT&T sees the activity/cost driver relationship as a means of getting to the root causes of environmental problems, whether they be product design, supplier qualification, and process efficiency on disposal of products at the end of their lives. AT&T's "Green Accounting Team" recognized that the goal of "customer satisfaction" may include customers other than the buyer of the manufactured product, such as society, shareholders,

[6]Jerry Krueze, "ABC and Life Cycle Costing for Environmental Expenditures," *Management Accounting*, volume 75, No 8, February 1994, p. 38.

the government, and even ecosystems. Doesn't all of this bears striking similarity to the large environmental management tools such as sustainable manufacturing and industrial ecology?[7]

How Might TCA and Activity Based Costing Conflict?

The key to effective ABC is the identification of cost drivers and allocation of costs across the activities. If these two steps are not performed with consistency and thoroughness, it will be difficult to extract the cost data needed to do TCA analyses. The previous section showed how TCA can be structured to reflect activities instead of cost categories. If, however, the activities needed for the TCA are not a part of the company's ABC program, then the same problem exists that plagues conventional accounting systems: The level of detail is simply not there. It is vitally important that you, as an environmental manager, be present and participate in the development of the company's ABC program. This ensures that the environmental activities are defined, their cost drivers discovered, and costs properly assigned to the responsible products. Without this foundation, TCA will be difficult to perform.

One company failed to have its accountants talk to the environmental management staff when ABC was being instituted. As a result, chemicals that had environmental impacts were grouped together with chemicals that did not, forcing the environmental managers to derive the data they needed for their decision-making processes and reports. This required developing a second cost tracking database and forced the environmental staff to continually update to maintain the usefulness of the database.

ISO 14000

This section answers the following questions:

- What is ISO 14000?
- How is it compatible with TCA?
- Where does it conflict with TCA?

What Is ISO 14000?

The International Standards Organization (ISO) is an independent standard-setting body with representatives from all industrialized nations. The

[7]U.S. EPA, *Green Accounting at AT&T.* Washington, DC: U.S. EPA, Office of Pollution Prevention and Toxics, September 1995.

organization's first major management standards, ISO 9000, focused on internationally recognized quality systems standards that required third-party certification. These standards met with great response from manufacturers because it provided a worldwide minimum quality standard. ISO 9000 is now almost ubiquitous in the manufacturing environment. In 1992, ISO expanded its focus to include standards for environmental management and called these ISO 14000. A team of representatives from 50 countries has been crafting these standards, which will eventually lead to internationally accepted environmental management practices. ISO 14000 standards contain six major components, each dealing with a different aspect of environmental management.

1. *Environmental Management Systems (14000 and 14001):* This segment states that a company must have a system in place for managing environmental responsibilities. The mandates require that a company have procedures in place for every environmental program action, from setting objectives, to maintaining records, conducting training, and auditing the performance of these procedures.
2. *Environmental Performance Evaluation (14031):* This segment mandates that an audit be conducted at some interval to inventory the facility's environmental impacts and identify the proper indicators of improvement. The audit does not actually rate the facility's improvement, just the presence of objectives, indicators, and procedures.
3. *Environmental Auditing (14010–12):* This standard outlines the procedures for evaluating whether the company's policy meets ISO standards. The auditor also would review the adequacy of procedures for communications, system documentation, and records control. In other words, are all the parts of the management system in place and maintained in the best judgement of the auditor? Some of these aspects require a high degree of knowledge of both the industrial processes and the national, state, and local environmental regulations in the area where the company is located.
4. *Life Cycle Assessment (14041–44):* Life Cycle Assessment (LCA) is a tool for determining a product's environmental impacts from creation to disposal. This section does not require a company to complete a life cycle analysis in order to obtain certification. Instead, it functions more as a resource base for a company if it should desire to conduct an LCA on their own.
5. *Environmental Labeling (14020–24):* Under this section, a company can self-declare certain environmental claims, use or create environmental

symbols, and self-test its claims. The section also includes common definitions of terms, and guiding principles and criteria for environmental certification programs.

6. *Environmental Product Standards (14060):* This section provides standards for other standards writers so that they may include environmental aspects in their product standards. It provides information on using tools such as LCA, design for the environment, and others to determine what makes a product environmentally preferable.[8]

The major attraction of ISO 14000 is the common ground it offers to ensure that certain key elements of environmental management are in place. It does not ensure compliance with environmental regulations; it ensures that there are procedures in place to help maintain compliance with environmental regulations.

How Does TCA Complement ISO 14000?

Surprisingly enough, amid the thousands of sheets pertaining to ISO 14000 requirements, there is little to no mention of cost accounting, environmental costs, contingent liabilities, or other factors so dominant in the world of environmental management. This is in part the result of compromises between all the countries involved, some of which desired extremely rigorous standards and others that wanted only a showpiece to hide behind. Still, it is unique given the amount of time (years) spent hammering out these standards that there is no discussion of environmental cost accounting or Total Cost Assessment.

This only means that there is perfect opportunity for integration. Because the standards are so loose, they allow companies to adopt a great range of practices that could enhance their Environmental Management System, Life Cycle Assessment, or Environmental Labeling Programs. TCA would make a great subcomponent to any company's Environmental Management System simply because the ISO 14000 standards provide no guidance on capital budgeting for environmental projects.

TCA can be used to assist businesses in choosing projects and reaching their objectives as defined by the Environmental Management System. Because the EMS specifies a declaration of environmental indicators (i.e., waste generation, water and air pollution, etc.), TCA also could be used to select

[8]CEEM, "We Have a DIS," *International Environmental Systems Update*, July 1995, p. SB-25.

options that reduce the impacts of these indicators. Environmental cost accounting systems also make great indicators and provide an integrated support system for TCA.

Where Do TCA and ISO 14000 Conflict?

Paragraph 3.14 of the Draft International Standard 14001 defines pollution prevention as "Use of processes, practices, materials or products that avoid, reduce, or control pollution, which may include recycling, treatment, process changes, control mechanisms, efficient use of resources and material substitution."[9] Most environmental professionals, especially state and federal regulators, would disagree that pollution prevention includes treatment and control of pollution. This statement may not appear to be in conflict with TCA, but it does point up a larger tone running through the standard that promotes a "do as little as you like" mentality. The lack of performance-based standards, the deemphasis on source reduction and substitution, the downplaying of the importance of Life Cycle Assessment, and the lack of any environmental cost accounting discussion all contribute to this pattern.

Because of this, within ISO-certified companies, there could be a disposition to "meet the status quo," or lowest common denominator. Strong cultural biases for conformity support this seemingly unreasonable hypothesis. Most companies and individuals do just what is needed to get by, rather than looking to alleviate future problems. Recall the theory of normal distribution, where the majority of subjects fall within a well-defined mean performance and smaller numbers of individuals fall on either end of the "bell curve." Environmental compliance is certainly no exception to this distribution, and if ISO 14000 certification becomes considered the mean level of performance, it follows that fewer companies will be willing to extend themselves beyond this established mean. Therefore, if TCA and environmental cost accounting are not included in the EMS, it is highly likely there will be no inclusion of it in the day-to-day environmental decision making. Companies will consider the cost of an environmental project; the sophistication of methods used may not be as high nor as accurate as it could be. As a result of this deficit, pollution prevention projects that could save the company consider capital and liability may not be pursued.

Consider also that public and corporate concerns for the environment have always run in a rather pendulumlike manner. Excitement and action

[9]Ibid., p. SB-29.

over environmental problems will be at a crest for several years and then will die down or even regress, depending on other factors like the national economy. It is possible therefore that EMS, while currently the rage, will fall out of favor in several years. Support for this speculation is garnered from the reactions of several plant managers who disclose no substantive gain from ISO 9000 certification. One in particular suggests it was a great way to spend a lot of money and generate a lot of paperwork.

LIFE CYCLE ANALYSIS AND TCA

Life Cycle Analysis is an emerging area that explores the impacts of products before and after the point of manufacturing. The impacts occurring before the manufacturing process might include the following:

- Extraction of the raw or base materials (i.e., copper ore, trees, etc.)
- Transport of these materials to a primary processing facility (by rail, truck, slurry line)
- Processing of these materials (smelting ore, refining, pulp processing)
- Synthesis of specialty materials and composites (carbon fibers)

Then traditional manufacturing of a product occurs, involving all stages of design, engineering, research and development, pilot testing and prototyping, tooling, production, assembly, marketing, packaging, and shipping.

There is yet another stream of activities following the product out of the facility doors that impact the environment:

- Shipping and transporting the product to retailers
- Discarding the packaging by the purchaser
- Using up the product's life, which may involve the combustion or consumption of nonrenewable materials
- Discarding the product once its life is deemed over
- Transporting the wasted product to a disposal site
- Disposing the wasted product

Each of the stages of this cycle creates an impact on the environment. These impacts may or may not have a direct cost to the manufacturer. The costs not borne by the manufacturer are borne by the user, society, and the ecosystems of the world. Some would argue that the cost of the product embodies the entire cost of all these activities. Unfortunately, that is inaccurate.

For example, the U.S. government subsidizes the extraction of minerals to both U.S. and foreign corporations. The true cost of mining a nonrenewable resource is not passed along to the consumer. Roads and highways are also subsidized through a myriad of policies and tax rebates. The cost of transporting oranges 3000 miles to your kitchen table is not reflected in the price of the orange because the trucker did not have to pay for constructing the roads the truck drives on and the price of fuel was kept artificially low through the use of military intervention overseas.

The theory of Life Cycle Assessment is to capture all the impacts that manufacturing a product creates. The next logical step is to apply a cost to these impacts. These costs and impacts then can be used in the design phases of production to select materials and manufacturing processes that reduce the impacts of the products. Theories such as Design for X allow a company to choose whether the product will be designed for recyclability, disassembly, minimum environmental impact, or all of these factors. See Figure 10.1.

The practice of LCA requires a well-delineated analysis structure. Think of the process as drawing a sphere around a product, with the goal being to

■ **FIGURE 10.1 Typical Life Cycle of Materials**

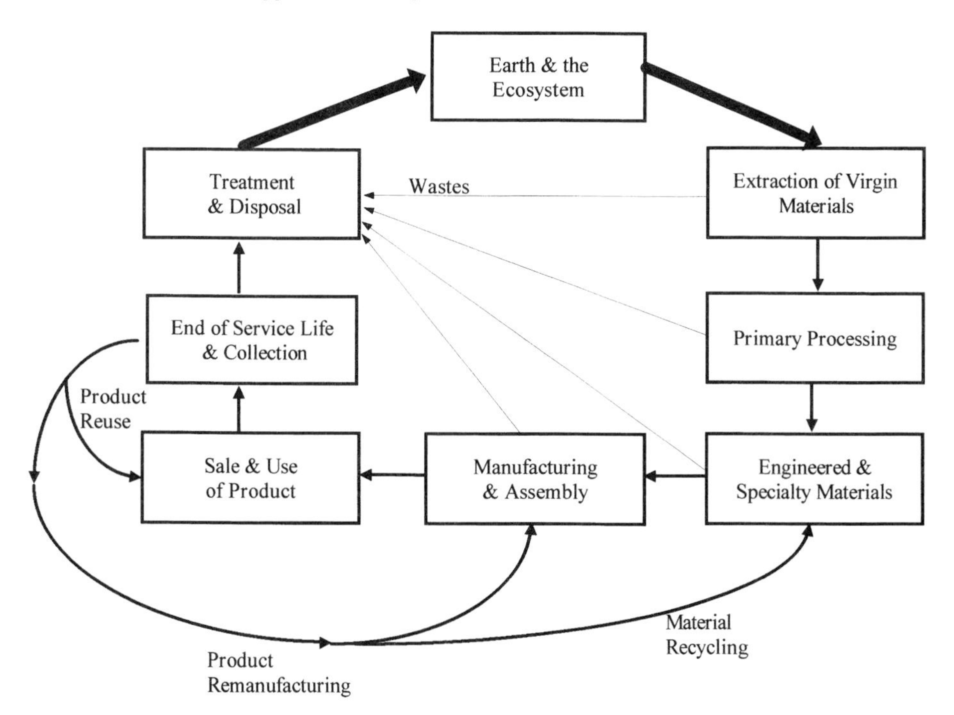

measure all impacts contained within that sphere. Obviously, the larger the radius of the sphere, the more impacts and costs will be incurred. However, the larger the sphere, the less likely the analysis will exclude a vital element or impact. LCA's strength is in its ability to address the entire materials use and disposal cycle and avoid cross-media pollutant transfer or pollution shifting.

How Is TCA Compatible with LCA?

The art or science of conducting a Life Cycle Assessment requires a suitable cost accounting tool that can extract costs for each impact whether it is a prior, materials extraction cost, or a future, end-of-product life disposal cost. No better tool exists than Total Cost Assessment. In fact, TCA is used in a somewhat expanded form in many LCA studies conducted in Europe and has been included in the U.S. EPA's *Life Cycle Design Guidance Manual.* The manual outlines a five-layer matrix of legal, cultural, cost, performance, and environmental considerations. Each layer is a matrix of requirements that organizes product, process, distribution, or management inputs and outputs of each requirement. Table 10.2 illustrates the matrix for environmental considerations.

To quote directly from the manual: "Modified accounting systems that fully reflect environmental costs and benefits are important to life cycle design. With more complete accounting many low-impact designs may show financial advantages."[10] Most Life Cycle Assessments will require a complex form of Total Cost Assessment. There may be 50 or more cost categories and a variety of inputs and outputs that must be tracked. The fundamental principles involved in analyzing the life cycle of a hydroelectric power plant located in an environmentally sensitive area are the same as for analyzing the feasibility of changing to a water-soluble metalworking coolant.

How Do TCA and LCA Conflict?

The major point of controversy in any Life Cycle Analysis is how big do you draw the sphere? At what point is it no longer feasible to study the impacts? To some degree, LCA suffers from the law of diminishing returns.

To illustrate, look at the creation of a cast iron frying pan. If just one frying pan is examined, the impacts of all the actions must be unitized to reflect the impact of that single frying pan. Remember, the analysis starts by defining a scope or a sphere of consideration around the object. For the first tier,

[10]Gregory, Keoleian, and Dan, Menary, *Life Cycle Design Guidance Manual*, Washington, DC: U.S. EPA, Office of Research and Development, January 1993, p. 50.

TABLE 10.2 Environmental Requirements Matrix for LCA

						Environmental	
	Raw Materials Aquisition	Bulk Processing	Engineered Materials Processing	Assembly & Manufacture	Use & Service	Retirement	Treatment & Disposal
Product							
Inputs							
Outputs							
Process							
Inputs							
Outputs							
Distribution							
Inputs							
Outputs							
Management							
Inputs							
Outputs							

assume the sphere extends no farther than the air around the frying pan–casting factory. All particulates and emissions from the factory must be unitized based on a weighted total production of all products, and one of these units is assigned to the fry pan, resulting in a fairly small number. Now extend the sphere out past the factory to the foundry, where the iron is refined from ore into ingots. Quantify the total impacts of this foundry, a large number no doubt. But only a unitized amount equal to the fry pan ingot can be included in the fry pan's LCA. This is probably a smaller amount than the first number. Now extend the sphere out past the foundry to the iron ore mine. What is the impact of open-pit iron mining? Quantify the total impact of all iron mined at the pit (probably a huge number) and determine how much iron ore it takes to make the ingot that makes the fry pan. Assign only this amount to the fry pan's TCA (probably a very tiny number).

Beginning to get the picture? When does the analysis stop? Should the fuel used by the cranes and drills that extract the ore be considered? What about the impacts of extracting the oil that is refined into the fuel that is used to mine the ore that is refined to make the ingot that is cast into the fry pan?

Now try to assign costs to these very small impact numbers. The costs are likely to be as small or smaller. Some calculators would not have enough place holders to count all the zeros.

To further illustrate consider the following real-life example taken from one industry's state pollution prevention plan. This plan required the quantification of costs associated with the use of a listed chemical and the calculation of the quantity of this chemical used per unit of product.

Chemical	***Amount Used in 1993***	***Amount of Product Shipped***
Manganese	104,685 lb	14,100,000 lb

At first glance, it appears that the company uses quite a bit of manganese (104,685 lb). The sight of such a large number out of context is analogous to the large numbers that would be seen from mining ore in the fry pan example. However, this amount must be unitized for the product being studied. Each product shipped in 1993 had only 0.00742 lb of manganese. This surprisingly low number is analogous to the unit impacts of mining ore that would be attributable to the single fry pan.

Next, the total cost of the listed chemical is determined. In the preceding instance, it was $60,719 for 1993. This represents manganese's share of all chemical purchase, waste management, regulatory compliance, insurance, production, maintenance, and utilities costs. This total is then distributed among all the products produced.

Total Cost of Manganese	***Amount of Product Shipped***	***Cost of Manganese/Unit***
$60,719/yr	14,100,000 lb/yr	$0.004306/unit

This is a fairly small cost. To the credit of most LCA researchers, the cost of a single fry pan is not the issue. The issue is the larger cost of producing all frying pans and whether it would be better to make them all out of aluminum rather than cast iron.

To summarize, TCA is placed at its numerical limits of accuracy when utilized this way. Researchers must be extremely careful with the transposition of numerals and decimal points. Without discrimination and thorough technique, the Total Cost Assessment could incorporate large inaccuracies.

TOTAL QUALITY-MANAGEMENT AND TCA

Total Quality Management (TQM) embodies a set of coordinated disciplines and management processes that ensure the customer's needs are always met or exceeded. TQM can be applied to a wide variety of organizational structures, from manufacturing plants to hospitals and schools. TQM engages all levels of the organization as team members and forces top management to organize its operations around customer satisfaction, employee participation, and continual improvement.

William Deming is often credited with creating the TQM movement in his work with post–World War II Japan. Rebuilding Japan required monumental coordinated efforts, which were driven by the country's desire to compete with the U.S. economy. Shunned by U.S. industrialists, Deming found an avid audience in Japan. The results are well known, as the flood of imports from Japan took over large portions of market share in many U.S. markets during the 1960s and 1970s. It was not until the late 1970s and early 1980s that Deming's models were employed at facilities in the United States.

TQM highlights five major principles needed for managing the organization and ensuring continuous improvement and customer satisfaction.

1. Manufacturers must be motivated by the threat of competition to find ways to exceed customer expectations.

■ FIGURE 10.2 The Shewhart/Deming Cycle

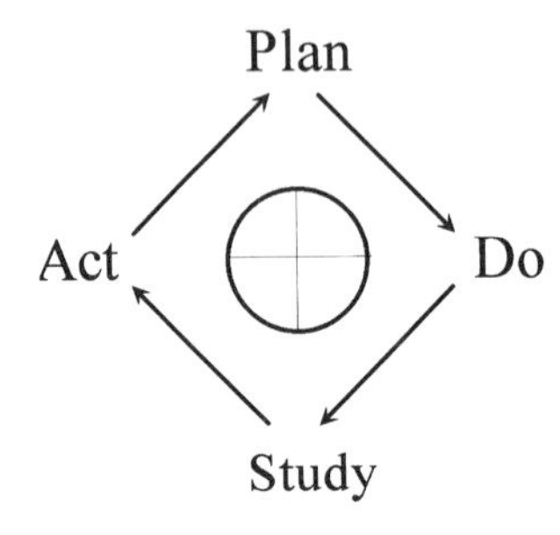

2. The management of the company must be fully committed to TQM at the highest levels and also make it clear that quality is everyone's responsibility. These require that employees be empowered to make changes or stop production if defects are spotted.
3. The focus must shift from a one-time major improvement to incremental changes that continuously improve the process.
4. Process controls and evaluation tools are used to design, implement, and evaluate processes and changes to them.
5. Set and adhere to a goal of zero defects. Achieve this goal through improvements to the front end or middle section of production rather than improving quality control inspections.

These principles are enacted through the Shewhart/Deming Cycle: Plan, Do, Study, and Act (see Figure 10.2).

Plan Planning is essential for smooth followthrough. Remember also that $1 spent in design saves $10 in production and $100 in warranty repairs.

Do Use tools such as the Ishikawa fishbone diagram, Pareto chart, trend chart, and others to record small-scale tests of the plan.

Study Review the results of the tests and make adjustments as needed.

Act Make a decision to either adopt or abandon the plan. You are now back at the top of the cycle to start over again.

How Does TCA Complement Total Quality Management?

As companies, or quality teams within companies, proceed through the TQM cycle of Plan, Do, Study, and Act, they will need tools to test the hypothesis and plans. They also will need also decision-making tools to provide feedback on the test results. TCA can play a role in each of these steps.

As a tool for testing hypotheses, TCA can be performed to determine the environmental cost of a particular quality improvement. For example, sometimes a quality improvement is at odds with worker health and safety. The specification for a certain level of cleanliness might predispose a company to use a highly toxic cleaning solvent. The quality team for the company might recognize this risk and decide to evaluate alternatives. Because workers can be considered customers, too, their satisfaction with the work environment would be important. Eliminating this chemical therefore may become a priority. TCA can be used to determine the cost impacts of the change and relative budget limits for a replacement.

TCA can be used again in the testing phase when the pilot tests from the "Do" phase begin to deliver cost data. These data can be synthesized through a TCA to provide the company decision makers with a financial perspective on the project.

There is also a great deal of overlap between the principles of TQM and the practices of pollution prevention. For example, both must have top-level management support if they are to succeed. Both try for zero defects, that is, no wasted product and no wastes from the product. Both emphasize a continuous approach to improvement and recognize that the first changes will be the easiest and require the least effort. Because of these similarities, TCA will fit quite well between TQM and pollution prevention. In fact, there is a growing movement for a Total Quality Environmental Management.[11] The example that follows describes some of the modifications that were performed to one company's quality assurance manual in order to integrate pollution prevention and environmental cost accounting.

The company desired to create awareness about environmental pollution as a product defect. One of the first modifications occurs in the executive summary pages:

> ________recognizes that environmental pollution and the generation of waste [are forms] of product defect. Therefore, ________ will include responsible environmental management and pollution prevention as a part of the quality program.

[11]PRC Environmental Management. *Total Quality Management/Pollution Prevention Integration Study for the Printed Circuit Board Manufacturing Industry*, Los Angeles: PRC EM, 1993, p. 13.

The company changed job descriptions and new employee training programs to reflect this new emphasis. Again, from the quality assurance manual:

> New employees shall be trained in the rules and regulations for safety. New employees will also be trained to recognize the importance of environmental protection, preventing pollution and why pollution is considered a quality defect.

Finally; the company also included changes to lot tracking procedures, cost allocation, and compliance inspections to the manual:

> The quality team shall review lot tracking records, waste generation records and materials purchase costs as one element of the improvement process. This review will be used to highlight areas of environmental defects and target future efforts.

Although it is clear that words in a quality assurance manual do not the practice on the floor make, it is nonetheless a step in the process of integrating environmental management with quality control.

Where Do TCA and Total Quality Management Conflict?

Total Quality Management stresses customer satisfaction. Depending on the breadth of the definition of *customer*, TQM can help or hinder environmental management. If the company focuses only on the customers that purchase the company's products, it is quite possible environmental issues will not be touched by the TQM process. There is a broader view that other types of customers exist, such as the community in which the facility is located, which expects a good neighbor, or the employees, which expect a clean, minimal-risk work environment, or the ecosystem, which it is hoped will continue to exist without risk of pollution or contamination. In this case, the company would be required to include factors affecting these customers into the TQM cycle. Total Cost Assessment is useful with the broad customer audience but not very useful for the narrow one. This creates conflict within the organization wherever there are costs that do not get incorporated into the product. In this case, the environmental costs might be excluded from review by the TQM team and eventually overwhelm the organization (i.e., in the form of a spill or site contamination event).

CHAPTER

11

LIMITS AND BARRIERS TO TCA

This chapter shows where the operational boundaries (limits) of TCA are and what barriers restrict TCA's effective application in other fields. The importance of knowing these boundaries cannot be stressed enough, which is to ensure that your TCA maintain the highest level of accuracy possible.

Strict dictionary definitions of "limit" and "barrier" differ only slightly and are somewhat circular in nature. For example, one dictionary states a *limit* is "the point, edge or line beyond which something cannot proceed,"[1] and a *barrier* is "something that limits or restricts."[2] For the purposes of this book and the concepts of TCA, there are distinct differences between a limit and a barrier. A limit to using TCA as a decision tool is not a function of failing to proceed, but more a substantial decrease in the reliability of the calculated values. In other words a *limit* is

> The point at which a system can no longer function, becomes unpredictable, inaccurate or unstable if exceeded[3]

[1]*American Heritage Dictionary.* New York: Dell, 1983, p. 398.

[2]Ibid., p. 57.

[3]Mitchell L., Kennedy, presentation to Tampa, Florida, businesses, November 12, 1996.

This definition clearly states that the limit is a condition beyond which use of the tool becomes inaccurate and likely to yield erroneous values. The TCA is constructed around gathering data and performing a net present value calculation, so the limits of these activities are also the governing factors of the TCA. The definition of a barrier will be presented after the discussion of TCA's limits.

UNDERLYING LIMITS OF DISCOUNTED CASH FLOW ANALYSES

As mentioned in Chapter 4, the practice of discounted cash flow analysis was originally introduced in 1958 in an attempt to bring mathematical rigor and a unified theory to capital budgeting and financial decision making. Prior to that date, many diverse approaches were used, most of which were more qualitative and intuitive than quantitative. At the time, these practices were accepted due to the vast number of uncontrollable variables in the financial marketplace.

In order to establish the "science" of financial decision making, Modigliani and Miller had to first establish controls or assumptions about the marketplace. These assumptions came to be known as the capital assets pricing model, or the paradigm of "equilibrium in perfect markets."[4] These basic seven assumptions laid the groundwork for all future financial decision making and the formulas used in the Total Cost Assessment. The assumptions are as follows:

1. All people having assets expect to maximize their gain in the given period of time and will make their decisions on the basis of the mean and variance of expected profits.
2. These people can borrow and lend an unlimited amount of money at a risk-free interest rate set by an external agent.
3. They have identical, subjective estimates of the means and variances of the expected profits on all assets.
4. The market for selling assets is perfectly competitive because no asset holder is buying assets.
5. There is a fixed number of assets and no new ones are issued.

[4]Micheal, Dempsey, "The Development of a Theory of Corporate Decision Making: An Historical Perspective with Implications for Future Development and Teaching." Ph.D. diss., School of Business and Economic Studies, University of Leeds, Leeds, England, 1996, p. 5.

6. The assets can be divided easily and sold easily with no transaction costs.
7. There are no taxes.

When all these assumptions are given, the price of the asset can be calculated as a function of the "risk-free" interest rate, the mean expected profit on the asset, and the variance of those expected profits on all the assets in the person's investment portfolio. Obviously the real world is vastly more complicated than the model. For example, everyone has estimates of assets' worth, there are always buyers in the marketplace who are also asset holders, new assets are issued all the time, and, of course, one of the two only constants in the world is taxes.

There is ample evidence to suggest that the model which all discounted cash flow analyses are based does not substantially represent the reality of the marketplace. Fortunately or not, most agree that it is the best model currently available and provides some level of consistency in the financial world. Just on speculation, how would you evaluate projects if you gathered all the cost data but did not have the net present value calculation with which to analyze them?

MATHEMATICAL LIMITS TO TCA

Data gathering and net present value analyses are fundamentally mathematical in nature, so the foremost limits to TCA are also mathematical. There are four situations to watch for when conducting TCAs, as they can introduce large inaccuracies in the calculations.

No Data Available

The reliability of the TCA decreases significantly in the rare instance where a facility has no physical data on a process or costs. Chapter 2 discussed the methods of estimating costs that are not directly tracked by the company. Most often, the largest costs of a process are well documented, leaving only the smaller costs unaccounted for. By their nature, these costs are proportionally smaller than known costs and an estimation of their value does not have as large an impact on total costs. Because of this, an estimation of these small costs that is off by even 100% would not have as large an impact as errors in the larger cost categories. If no cost data are available, the amount of error introduced into the total cost is magnified. The value of conducting the TCA on such a process must be questioned for two reasons. First, the results will be questionable. Second, a larger than normal amount of time will be required to conduct the TCA because all the data will have to estimated.

Monolithic Costs

Large inputs of one particular material may create material input costs 100, 1000, or 10,000 times larger than any other cost category. The shear size of these costs impacts the profitability of any change that does not affect the usage of these materials. For example, usage of sodium hydroxide at a kraft bleaching mill costs $230,000/yr and the spare parts for maintaining the pumps and mixers that churn the pulp costs $50/yr. A TCA performed on replacing the pumps and mixer seals more frequently to decrease spills and leaks of bleach solution would not show appreciable changes in operating costs for the kraft bleaching process because the chemical purchase costs are so much larger than the pump seal replacement costs. It may not be possible to ignore these large costs when performing the TCA, resulting in a deceptively flat financial performance of the alternative.

Projects That Never End

Caution must be exercised when analyzing very long-term projects. A very long-term project can be considered anything over 15 years.While they never used discounted cash flow analysis, many Native American tribes considered the repercussions of their actions out to the seventh generation of their children when deciding tribal policies. Planning for that length of time requires conservative thinking, high-quality data, and a minimum number of initial assumptions. Changes in the national economy, rates of technological change, and factors external to the company are more likely to interfere the farther the project timeline extends into the future. The net present value calculations should be augmented with factors representing changes in market prices for goods, and the impacts of activities outside the company may need to be considered.

Extremely Expensive Projects

As mentioned in Chapter 4, the compound interest tables have a limit to their accuracy due to the number of significant digits in the printed columns. This will never present issues for the majority of projects being analyzed. For very large project costs (i.e., over $10 million) the lack of significant digits creates a serious problem. In these cases, tens or even hundreds of thousands of dollars can be lost due to rounding errors inherent in these tables.

One method to overcome this limit may be to manually calculate the present values using a computer or calculator with scientific notation.

SOFT-APPROACH POLLUTION PREVENTION AND PERCEIVED TCA LIMITS

You may find most of your TCAs conducted on the financial benefits of *technological* changes to production processes. Another method to achieve cost savings and prevent pollution deals primarily in retraining workers so their practices are more efficient and less wasteful. These techniques are often called the *soft approach*. Training may include how to notice opportunities for improvement, how to handle materials and products more carefully, and how to be more conscious of the use of materials. Examples of this soft approach to pollution prevention, include the following:

- Training forklift operators not to skewer drums, cartons, and so forth
- Making chemical process baths using documented formulas and measurements
- Holding parts over baths longer to drain off excess solution
- Adjusting water pressure manually to decrease usage as production slows down
- Segregating wastes by placing materials in separate containers
- Using dry cleanup techniques to conserve water

These soft approaches entail more than just implementation and cost tracking. They also require retraining employees, extended management commitment, and visible display of this commitment through daily supervision, awards, consequences, or other means.

TCA can be used to evaluate both the change to a new technology and the so-called soft approaches. It is, easier however, to study inanimate objects and chemical formulation changes than to try and account for the variety of interactions that effect the human condition out on the shop floor. Data gathering for soft-approach projects takes more of an observational quality than a strict review of files and records.

Robert J. Sternberg's *In Search of the Human Mind* states that among the many research methods available to psychologists, naturalistic observation yields the best understanding of behavior in natural conditions and the results tend to be widely applicable. Yet the lack of experimental control and likelihood that the presence of the researcher will influence the observed results works against this method.[5] To illustrate these difficulties consider the following scenario.

The shop foreman at a manufacturer of plumbing fixtures desired to increase production in the polishing department; he went down to the process to observe how the fixtures were polished. Each plated fixture is hand-buffed on rotating drums and the workers in this line are paid based on the number of parts buffed per hour, with a bonus given after a certain number of parts are completed. The line workers learned of the manager's coming inspection and purposefully slowed their work rates to deceive the manager. When the manager timed the workers to develop a new quota, the rates did not fit with the throughput of parts processed by departments preceding and following the buffing line. Later that day, the manager returned to the polishing department and hid among the crates of polished fixtures and timed the buffing-line workers. This time he obtained a more accurate piece rate and was able to adjust the production figures.

Workers generally know what they *should* be doing; however, they may not do it for any number of reasons as long as they feel they will not be caught. In order to gather accurate data for a TCA involving primarily soft costs, you must eliminate the confounding variables that bias the data gathering.

For example, an electroplating job shop has several plating lines referred to as "hand lines." The line workers raise and lower the plating racks into the tanks by hand without the aid of mechanical hoists. The racks of parts may weigh anywhere from 10 to 60 lb. Hand lines typically lose more plating solution than lines where the racks are raised and lowered by a mechanical hoist. Table 11.1 shows a comparison between the variables affecting solution loss from hand lines versus automatic plating lines.

The table contains a partial list of confounding variables, some of which can be controlled or eliminated (i.e., lighting, room temperature, rack weight) and others that cannot be controlled (i.e., disposition of workers toward management, worker's mood, and energy level).

If consistent and minimized solution loss is the goal, several soft P2 approaches could be tried, such as encouraging the workers to hold the racks over the tanks longer, pull the racks out of the tank slower, or using a spray wand to rinse the rack over the plating tank instead of over the rinse tank. The potential benefits from these changes can be measured theoretically by time-and-motion studies of the workers' practices and sampling of plating solution

[5]Robert J., Sternberg, *In Search of the Human Mind.* Orlando, FL: Harcourt Brace, 1995.

■ TABLE 11.1 Comparison of Factors Affecting Plating Solution Dragout on Two Lines

Hand Line	Automatic Line
Energy level of the line workers	Amount of preset dwell time over the tank
Workers' moods on that day	Jammed or worn hoist parts
Disposition toward management	Pre-set hoist rise rate
Lighting conditions	Voltage fluctuation
Weight of the racks	
Work flow rates and quotas	
Temperature of the plating room	
Number of breaks allowed	

loss. The question is not whether TCA can quantify the benefits, but how accurate or reliable the results will be. The computer programmers' axiom, "garbage in, garbage out," applies to the reliability of data gathered for "soft pollution prevention."

Projects that seek to change worker habits, attitudes, or tasks always require more management involvement, either direct oversight or incentive awards. The results of the project depend more on the ability of the company management to effect lasting change in the work place, than the ability of TCA to quantify the benefits. The less management is involved in implementing the project, the less successful the soft approach will be regardless of the accuracy of the TCA. The TCA is not in question in these cases. The degree to which the TCA of a soft P2 approach appears to be inaccurate is a reflection of the inability of management to effect change within the organization.

Conversely, equipment replacement projects yield relatively consistent pollution prevention results due to the mechanization of production. In equipment replacement projects, the accuracy of the TCA simply reflects the predictability of the investment in equipment. Human free will is less predictable. The TCA is simply a tool, not the change agent. You cannot perform a TCA and expect its results to miraculously change the way people work.

Barriers to TCA

Whereas a limit is a distinct point at which the TCA calculations become unreliable, a barrier has a completely different connotation. A *barrier* is

> A hindrance or obstruction preventing continued motion or progress, particularly if the objects' limits of performance have not been reached.[6]

In this book, a barrier is something external to the TCA process that hinders the completion of the TCA or application of the TCA methods to the situation. In this case, it may be possible theoretically to conduct a TCA, yet some regulation, policy, or technological obstacle prevents this from occurring.

Understanding the difference between a limit and a barrier is vital. Exceeding limits results in unreliable values, whereas overcoming barriers allows new situations to be analyzed using the TCA method. There are two major categories of barriers:

- *Internal barriers* originate from within the corporation's structure, policies, or its employees. You have some degree of control over these barriers and can act as a change agent to remove them or minimize their damage to the TCA and environmental projects.
- *External barriers* originate from outside the corporation's structure and result from macroeconomic policies or cultural biases. Your control over these barriers can extend only to edge of the factory, and only that far if you fully understand how they impact the operations within the facility.

CONFLICTS WITH THE CURRENT ACCOUNTING SYSTEM (INTERNAL BARRIER)

Because you are functioning within a facility and an established business culture, the current cost accounting system must be considered as a potential barrier. The corporate cost accounting system may be too "big picture" to provide useful data, or too restrictive in the type of data collected. Particularly in large, multinational corporations, cost accounting systems can be completely different from one facility to the next. This would interfere with the transfer of cost data between facilities or the gathering of data from a central office. One multinational corporation had such diverse cost accounting systems between facilities that corporate initiatives were stymied because no two facilities had the same indicators of performance!

Another potential barrier may be the overwhelming amount of processes and products produced by the company. Total Cost Assessment selectively

[6]Mitchell L. Kennedy, presentation to Tampa, Florida, businesses, November 12, 1996.

isolates operating costs for particular processes or products. It is feasible theoretically to identify, track, and document every cost and update all of this information every, week, month, or year. However, this could be a lifelong task in and by itself.

Develop a Cost Accounting Metric

One possible mechanism to deal with too many costs and products to track all the time is to develop a conversion metric for each product or process based on accurate historical figures and an adjustment that offsets changes in production flow or materials costs. This way the costs would be established once and adjusted up or down to coincide with fluctuations in production, without having to check the cost of every process.

This does require an initial effort to establish the base costs, but this only needs to be done once. A random sample of the costs could be selected for updating every month or year to maintain the accuracy of the system, but would not require anywhere near the labor of updating the entire system with equal frequency.

Document All Assumptions

Another key to the successful translation of calculations among specific projects, industries, or corporate divisions is defining all assumptions and estimates at the outset. The assumptions then can be validated or rejected and adjustments can be made to each project. Coefficients and conversion factors can be used to change the apples to oranges.

Here are a few guidelines for documenting any assumptions made due to estimating missing data or comparing dissimilar data.

- Lack of receipts, bills, and records will force an omission of that element or the estimation of its cost.
- The fewer number of costs entered, the more conservative the net present value.
- All omitted elements should be stated in the concluding remarks.
- Using estimates decreases accuracy proportional to the dollar amount estimated.
- Choosing the minimum value from a range will reduce the total inaccuracy, but numerous estimated values will compound the inaccuracy.
- All assumptions and estimates should be recorded in the summary to substantiate the final number.

LACK OF LONG-TERM PLANNING (INTERNAL BARRIER)

Another limit to the effectiveness of TCA is management's willingness to consider long-term costs and benefits. The TCA's ability to refine and summarize cost savings over a project lifetime is not useful if management will not consider planning beyond the next fiscal year. This limitation also strikes the heart of pollution prevention, as many P2 projects have longer paybacks when compared to competing capital purchases. This limitation can be removed only through change in management opinion and a desire to plan for future events.

RESISTANCE TO CHANGE (INTERNAL BARRIER)

In your role as environmental manager/cause champion/salesperson extraordinaire, the most valuable experience you can have is in dealing with how others handle change. One of the most common barriers to any project is human's fundamental resistance to change. Perhaps in this day of increasing technological innovation, change is more uncomfortable than ever. W.C. Fields once said:

> The only person who likes change is a wet baby.

This quote would not be humorous if it did not ring true to most people's experiences. As a corporate professional and part-time salesperson of pollution prevention projects, you must recognize this resistance for what it is: nothing more than an initial reaction and request for more information. Any course in salesmanship will explain that the words "I'm not interested in your_______product" actually mean "You have not given me enough information, or the right type, to pique my interest." Always keep this in mind as you present new concepts or projects to management.

Change management consultants have described four distinct levels of resistance to change, each requiring its own specific approach to overcoming it. These levels are described in what follows.

First Layer of Resistance

"We don't need it—everything is fine." Classic psychology might call this stage *denial*, because it rejects the possibility that anything could be wrong or in need of improvement. At this first stage, people will tell you there is nothing wrong with the current system; that it has always been this way; if it ain't broke, don't fix it; or other excuses to prevent change. The excuse may originate from not seeing the forest for the trees and therefore not seeing problems or possible improvements. If your audience feels threatened by the change, this will also form part of the first resistance.

Second Layer of Resistance

"There is just no way to make the change." This layer shows an acceptance of the need for change but resignation that things never change. "We tried that before and it failed" "We could never complete anything like that." These disempowered statements can be met with overt optimism and solid facts about the benefits of the project and the other successes.

Third Layer of Resistance

"Yes, but all these other issues will cause it to fail." Here the objector has conceded that there is a need for a change, and that your idea is feasible, but other factors will prevent it from working. It is much easier to find fault with new concepts than to find what is worthy in them, so this layer can be particularly thick. The excuses at this level include: "It will create too much paperwork." "It's not in the budget". "We don't have the labor force." "We can't compromise quality." Try to anticipate these questions and have answers ready. Stress the benefits and savings of the project, especially those that directly impact the resistant person.

Fourth Layer of Resistance

"Okay I'm in, but the_______[customer, manager, president, stockholders] will never go for it." Here your former adversary is on your side and suspects everyone else will be against the two of you. At this point, the seeds of resistance may be replanted and the process might occur over again, but only now there is one additional person on your side.

Peeling away of these layers of resistance creates buy-in to the project. Giving up at any one of these layers can mean the end of the project. For additional information on getting your ideas approved and overcoming these tired old excuses, read Charles Thompson's book, *"Yes, But . . . "—The Top 40 Killer Phrases and How You Can Fight Them.*

THE EPHEMERAL NATURE OF SAVINGS (EXTERNAL BARRIER)

The main strength of TCA is its ability to quantify savings resulting from a pollution prevention project. These savings traditionally have been undiscovered, allowing less environmentally beneficial projects to win out. Businesses perceive cash flows in a narrow window of time, usually between present day and two years hence. A recent U.S. EPA survey reported that 60% of the respondents required a payback in 2 years or less and 80% required the payback in 4 years or less.[7] The saying " What have you done for me lately?" is quite appropriate.

This myopia creates a barrier to appreciation of TCA's longer-term benefits. Although the savings are often significant and last for the entire lifetime of the project (5, 10, 15 years), the savings usually are lost in the immediacy of daily activities. This results in the "ephemeral" nature of savings, illustrated in what follows.

A facility conducts a TCA on changing metalworking coolants from straight petroleum oils to vegetable-based oils with microapplicators. The benefits of this change include the following:

- Decreased purchase costs for machine coolants
- Elimination of potential liability for oil spills
- Elimination of compliance costs for spill containment and control
- Decreases in labor for maintenance of coolant delivery systems
- Reduction in tool breakage
- Reduction in heat stress of parts being worked
- Elimination of oil mist in plant air
- Decreased need to clean oils on the outside of machines
- Potential to run at higher speeds and feeds, thus higher throughput rates
- Decreased waste oil disposal costs
- Possible part quality improvements
- Less tool breakage, an ancillary operating cost savings

This list shows impressive potential savings. Once the project is implemented and the new coolant works, these savings will begin to accrue.

An Example of the Phenomenon

One shop that performed such a changeover realized a $160,000 savings in one year. Of course, no one gave the company $160,000; the company just did not have to pay out the money in labor, parts, chemicals, and waste-disposal costs. In the first year, everyone felt great about the project and its success. The company enjoyed a similar cost savings in the second year, because the operating costs for the facility stayed roughly the same as the previous

[7]U.S. EPA, *Environmental Cost Accounting for Capital Budgeting: A Benchmark Survey of Management Accountants*, Document EPA742-R-95-005 Washington, DC: September 1995, p. 42.

year. Financial data for the year before the changeover were also readily available and they cross checked the before and after numbers. In the third year, the operating costs increased in step with economic inflation and the company turned its focus to other issues, feeling that this project had come to completion. Comparing the third year to the previous two does not emphasize the large cost savings achieved 3 years ago.

Because of the lack of visibility of this savings in the cost accounting reports, the savings became taken for granted. Of course, the project continued to save the company over $160,000/yr, perhaps even more if future prices of petroleum-based cutting oils increased faster than vegetable-based cutting oils. These additional savings occur but are not tracked because the company no longer uses petroleum-based oils. This apparent disappearance of the savings is exactly what is meant by the ephemeral nature of savings and can be considered something of a phenomenon.

The Phenomenon Continues

A second, more personal example may help to further clarify this issue. Suppose you own a new home and heat your house and hot water with natural gas. Annual natural gas bills average $2,000/yr because you live in a northern climate with cold winters and your spouse does not like to be cold.

You see an advertisement in the paper for passive solar space and water heating and decide to check it out. The TCA you perform indicates a profitable return on investment. In fact, it could reduce the monthly bills considerably and create a little more room in the household operating budget. You and your spouse decide to switch to passive solar heating, resulting in $1,500 savings in your natural gas bill. You know this because the operating costs for your house have dropped. In the first year, you are both ecstatic at how smart it was to invest in solar heating. So ecstatic in fact, the savings is quickly spent on a short vacation to the Caribbean. The second year you still feel smug that you get three-fourths of your home heating free from the sun and decide the "extra" money can go to some other projects around the house. In the third year, the solar heater is still saving you money but it is not visible in the operating cost for two reasons:

1. Comparing current operating costs with those from the previous year shows no cost savings.
2. The operating budget stayed at the same level, allowing other projects to use up the cash formerly paid to the natural gas utility. This allowed you to do more with the money, but maintained the same level of cash outflow from the household.

The Cultural Bias against Savings

The issue just illustrated is not unique to the fields of either TCA or pollution prevention. It is in fact a fundamental barrier not just to TCA, but to many other larger issues and even national budgetary problems. As a culture, the United Sates has lost its appreciation for long-term savings by focusing on immediate financial returns and the drive to spend as much capital as is available.

Such has not always been the case. During the 1930s and 1940s, the people of the United States developed an acute "conservation ethic" as capital became scarce during the Depression and the resources were often rationed during World War II. Since then, the ethic has been eroded by a drive to increase the standard of living at all costs. Juliet Schor described this in her book *The Overworked American* where she details how the per capita number of hours worked per week over the past 50 years has increased despite the introduction of scores of "labor saving" devices. She ascribes this paradox to the people of the United States valuing the acquisition of material goods over time off from work.[8] It is beyond the scope of this book to attempt to rectify a national cultural bias, however, there is one method you may be able to use within a company to begin building appreciation for savings and pollution prevention. It is best illustrated in the following example.

How to Capture the Elusive Savings

A manufacturer of consumer hand tools took an innovative approach toward valuing pollution prevention and waste reduction projects. The company has 150 full-time employees and is privately owned. The purchasing manager learned of pollution prevention through the state technical assistance program. He initiated the first project on a wager with the president that he could save the company money. He chose a small, easily accomplished task and won the favor of management. After several additional successes, the purchasing manager desired to create a budget for purchasing environmental products. The company agreed to allocate his department a portion of the savings from each project. In fact, the company split the savings among the owners of the company, the employees, and the newly created budget for environmental projects. Each successive project has increased his environmental budget, allowing the purchasing manager to increase the size and scope of the next project. This program saves over $500,000 annually and now involves a full environmental management team actively searching for ways to reduce pollution.

[8]Juliet Schorr, *The Overworked American*. Boston: HarperCollins 1991, p. 123.

The preceding example illustrates the principles of reinvestment, small victories, and continuous improvement. By developing a strategy to capture the savings and consciously reinvest it into visible budget-line items, the company successfully retained a perspective of how much money is being saved from year to year. Secondly, the reinvestment of the capital reinvigorates the workers and provides incentives for additional improvements. Note that the first several successes did not fall under the reinvestment strategy, but served to build excitement about the potential of pollution prevention. The company did initiate continuous improvement program midway through the development of the environmental group. The tracking of costs associated with a particular process helped raise awareness about the environmental impacts of the processes.

THE EFFECTS OF THE MACROECONOMY (EXTERNAL BARRIER)

Just as your functions as environmental manager are inseparable from the operations of the company, so, too, are the functions of the company inseparable from national and international economic systems. By understanding how the larger systems affect the smaller ones, you will be better able to anticipate the future environmental needs of the company, which technologies will be useful in meeting those needs, and how best to use TCA in evaluating them.

A corporation can be thought of as a small ecosystem or operational unit. As such, each is unique in the way management runs the company, the type of materials used, and the accounting system used to track production costs. However, interactions with entities outside the company require some degree of standardized method of conducting business. Within the United States, the federal government prescribes which methods will be used to measure economic activity (this is effectively the costs and profits of national commerce).

It has been well documented that our macroeconomic system does not assign economic value to pollution, depletion of resources, and future liabilities. Because of this, there is a disconnect at the door of the industry between the ability to use Total Cost Assessment within its walls to track costs and the need to sell materials and services to the rest of the world.

External costs for mineral and timber extraction do not reflect the impact of strip mining and clear-cutting on a national level, so the retail price of goods from a specific company practicing TCA cannot competitively reflect these costs. The first company to try to incorporate the larger costs of global or regional environmental damage into its product will be at a competitive disadvantage because the other companies will not participate. Or they may participate initially and cut prices if market share decreases. Consumers have been shown to choose low cost over high quality time and again. The success of large discount stores are an example of this principle. Research into green marketing shows that at best 10% of the populace consciously selects products based on their environmental impact despite increased costs. Others may desire to purchase environmentally friendly or recycled goods but cannot afford the cost difference. This does not provide enough incentive for companies to incorporate these costs.

TCA IS TOOL IN THE ARSENAL RATHER THAN A SILVER BULLET

Many times, there is a tendency among professionals in the environmental community, and elsewhere, to seek a "magic silver bullet" that will transform the landscape and solve all issues. This overly simplistic view results in disillusionment with techniques and their premature demise. Total Cost Assessment will not create the closed-loop zero-discharge industry. People within the organization must be committed to that goal and use many other tools to achieve it . TCA is a tool for showing what is traditionally hidden, yet intuitively known: that sound environmental practices actually save capital and decrease operating costs. It is vital to use all the tools to create the solution. You cannot build a house with just a saw.

CHAPTER 12

"BEYOND THE WALLS OF THE FACTORY"

Now that you have mastered the concepts of TCA, we will look at two examples of how it is used to analyze environmental problems on a global level. Questions are often raised: Can the methods of TCA be used in any other applications? Why or why not?

SIMILARITIES BETWEEN THE WORLD AND THE FACTORY

The operations of a single corporation can be compared to the operations of a nation or the world. The similarities include the following:

- The evaluation of solutions to environmental problems that entail risk to either the safety of employees or the well being of the citizens.
- The flow of materials into and out of their boundaries, which are either the walls of a factory or national borders.
- The collection of data to define the total impact materials use has on the environment in terms of dollar value, products manufactured, or goods exported.
- The allocation of pollution costs to the overhead category of national budgets, just like the operational budgets of most companies.

THE UN MODEL OF GLOBAL TCA

In 1990, the United Nations Conference on Environment and Development (the Earth Summit) created a proposal in Agenda 21 to integrate environmental and economic accounting methods in the System of National Accounts (SNA).[1] Whereas many countries have developed their own environmental accounting processes, Japan is the only country to have adopted the revised SNA.

JAPAN'S EDP

In 1995, Japan's Economic Planning Agency (EPA) completed a trial research project to estimate the environmental impact of economic activities and relate them to the Gross Domestic Product.

The environmentally adjusted Net Domestic Product (Eco-Domestic Product, or EDP) is derived by subtracting the environmental costs from the Net Domestic Product. This charges the cost of environmental degradation and resource depletion against the economic gains seen during the same period. This formula has the beneficial result of demonstrating the loss of natural resources on the total value of the nation's wealth. The Japanese did not account for all environmental costs, but chose to analyze those areas most important to their society.

Cost data were collected for the following pollution generating activities:

- Air pollution: SO_x, NO_x
- Water pollution: biological oxygen demand (BOD), chemical oxygen demand (COD)
- Waste disposal
- Destruction of the ecosystem by land development and deforestation
- Extraction of subsoil resources: coal, ore, etc.

Japan's national GDP for 1990 was 425 trillion yen (approximately $4 trillion). The study found that the EDP accounted for 5% of the GDP, or roughly $200 billion, without the inclusion of many other significant environmental impacts. The next generation of the EDP will attempt to expand the scope of estimation to include the following:

[1]The concept and structure of such accounting is shown in Chapter 21 of the revised SNA manual, as well as in a separately published *Handbook of National Accounting: Integrated Environmental and Economic Accounting.*

- The environmental costs of deforestation and resources lost in foreign countries that supply Japan's imports
- Other air-polluting substances, besides NO_x and SO_x (e.g., nonmethane hydrocarbon, CO, and floating particles)
- Other water pollutants, besides BOD and COD (e.g., nitrogen and phosphorus)
- Soil pollution, noise and vibration pollution, and the maintenance cost of water resources
- Global warming, destruction of the ozone layer, and other global environmental problems (e.g., CO_2, fluorocarbon gas, methane gas, nitrogenoxides)
- The valuation of scenic beauty, plants and animal life, and other amenities
- Positive valuation of environmental assets, including the environmental protection effect of forests and farms.[2]

THE ECOLOGICAL FOOTPRINT APPROACH TO GLOBAL TCA

Another approach to evaluating the environmental impacts outside the industrial complex is Mathis Wackernagel and William Rees' methodology called the Ecological Footprint. The method evolved as a teaching tool that links land use planning, architecture, and environmental issues. In trying to convey the linkages and cause-and-effect relationships of planning decisions, the authors found a versatile technique for showing the impact of a single person, village, region, or country.

WHAT IS AN ECOLOGICAL FOOTPRINT?

Wackernagel and Rees propose a nonmonetary system for indicating human ecological impact on earth. Their system involves quantifying the land used to produce and dispose of all materials consumed, and the activities conducted by a person, city, region, or country. The data required for such an analysis

[2]Noboru Nishifuji, *Trial Estimates of Integrated Environmental and Economic Accounting: The Report of Research Study to Add the Environmental and Economic Accounting to the System of National Accounts*, Japanese Global Environment Research Fund of the National Environmental Agency, the Economic Planning Agency, Keio University, and the Japan Research Institute, 1995.

are freely available through federal databases and environmental texts. The Ecological Footprint then can be compared to a theoretical average allocated space (3.7 acres per person—derived by dividing the total livable land by the total human population). The degree to which a person, city, region, or country has more acreage per person than the average indicates the amount of conservation required to be sustainable.

CAN ECOSYSTEMS BE ACCURATELY PRICED?

Wackernagel and Rees take the approach that the human impact on the ecosystem cannot be valued accurately. They argue that monetary analyses of natural resource use and sustainability are flawed for the following reasons.

1. Placing a monetary value on natural resource supplies hides the finite nature of the resource and its depletion. Traditional macroeconomic theory states that decreased supply should be reflected in increased prices. This would allow a consistent income and creates the illusion of a constant supply of the item. Conversely, improvements in harvesting technologies allow for the rapid depletion of remaining resources, causing market prices to fall and maintaining the illusion of steady supply (or abundance) of the resource.
2. Market prices for goods do not provide information about how much is left or when it might run out. They do not show critical limits beyond which irreparable harm is incurred. The authors give the depletion of fishing stock in the Atlantic Ocean as an example of technology allowing prices to remain constant while the resource (fish) becomes depleted. Most resources that could recover from critical overuse (i.e. deer, bison, fish, etc.) require several generations to do so, creating a time lag that is also not reflected in the price.
3. The intrinsic value of nature remains constant or perhaps increases with time, whereas the value of money decreases with time. This conflict results as fixed resources and ecosystems are forced into our growth-oriented economic structure. Because the ecological system cannot grow or expand in the same limitless fashion as our economies, there will be the appearance of a decline in monetary value for all resources. These resources will continue to be as vital as they are today (i.e., we will still need air to breath), but would be discounted through the time-value-of-money theory.
4. Markets cannot represent the value of many life-support processes such as nitrogen fixation, global heat distribution, and climatic stability. These processes are larger and more complex than economic systems, having impacts

on multiple species, continents, and other processes that may be unknown. For example, the loss of ozone over the equator is now suspected to be the cause of a die-off in coral reef organisms. This impact had not been anticipated previously and may have severe implications for tropical marine ecosystems. The taxes placed on chloroflourocarbon products could not account for the damage caused by these reactions.[3]

EARTH-CENTERED VERSUS ECONOMY-CENTERED PERSPECTIVES

Ecological Footprints answer the questions: "How far beyond the average land and resource use, have we (individuals, cities, regions, or countries) extended ourselves? And how much should we reign in our future consumption?"

This contrasts with the viewpoint of the Japanese project that answered the question: "How much does environmental protection and resource use impact our national economy?" Think of this as the "economy-centered viewpoint," where the environment is another factor tallied into the economic performance of the nation. Because the footprint approach excludes monetary analysis completely, it is more difficult to answer the questions: "How much will the required changes cost me (the consumer)?" "Can we (society) afford to do it all now?"

The Japanese approach is strong in the dollar value category, but lacks a critical assessment of how bad the ecological damage is. Presumably, as earth's environment degraded, this method would show economic impacts increasing directly or exponentially.

SUMMARY

TCA is a tool best used by the unrelenting detective, discovering cost data, tracking through forgotten records, and presenting the astonishing findings. The framework used to establish costs and determine the value of one project over another can be used in countless applications. It is at this point that the science of accounting is blended with the art of investigation. Being curious and able to look at processes and products from different perspectives are prime requirements for becoming practiced in TCA, whether on the facility level or on the global level.

[3]Mathis Wackernagel and William Rees, *Our Ecological Footprint—Reducing Human Impact on the Earth.* British Columbia: New Society Publishers, 1996.

APPENDIX

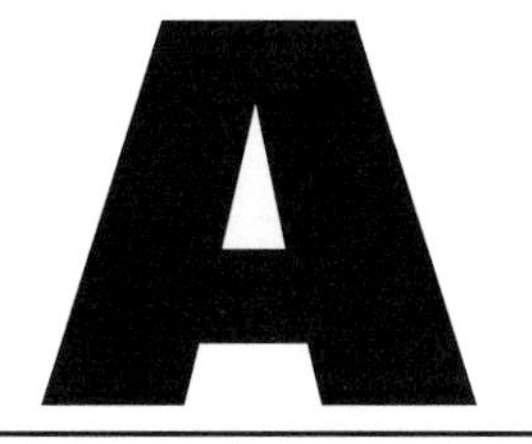

COMPOUND INTEREST TABLES: PRESENT VALUE OF AN ANNUITY (3%–22%)

Number of Payments	3%	5%	7%	9%	10%	12%
1	0.9709	0.9524	0.9346	0.9174	0.9091	0.8929
2	1.9135	1.8594	1.808	1.7591	1.7355	1.6901
3	2.8286	2.7232	2.6243	2.5313	2.4869	2.4018
4	3.7171	3.546	3.3872	3.2397	3.1699	3.0373
5	4.5797	4.3295	4.1002	3.8897	3.7908	3.6048
6	5.4172	5.0757	4.7665	4.4859	4.3553	4.1114
7	6.2303	5.7864	5.3893	5.033	4.8684	4.5638
8	7.0197	6.4632	5.9713	5.5348	5.3349	4.9676
9	7.7861	7.1078	6.5152	5.9952	5.759	5.3282
10	8.5302	7.7217	7.0236	6.4177	6.1446	5.6502
11	9.2526	8.3064	7.4987	6.8052	6.4951	5.9377
12	9.954	8.8633	7.9427	7.1607	6.8137	6.1944
13	10.635	9.3936	8.3577	7.4869	7.1034	6.4235
14	11.2961	9.8986	8.7455	7.7862	7.3667	6.6282
15	11.9379	10.3797	9.1079	8.0607	7.6061	6.8109
16	12.5611	10.8378	9.4466	8.3126	7.8237	6.974
17	13.1661	11.2741	9.7632	8.5436	8.0216	7.1196
18	13.7535	11.6896	10.0591	8.7556	8.2014	7.2497
19	14.3238	12.0853	10.3356	8.9501	8.3649	7.3658
20	14.8775	12.4622	10.594	9.1285	8.5136	7.4694
25	17.4131	14.0939	11.6536	9.8226	9.077	7.8431
30	19.6004	15.3725	12.409	10.2737	9.4269	8.0552
40	23.1148	17.1591	13.3317	10.7574	9.7791	8.2438
50	25.7298	18.2559	13.8007	10.9617	9.9148	8.3045
75	29.7018	19.485	14.1964	11.0938	9.9921	8.3316
100	31.5989	19.8479	14.2693	11.1091	9.9993	8.3332

(Continued)

(Continued)

Number of Payments	14%	16%	18%	20%	22%
1	0.8772	0.8621	0.8475	0.8333	0.8197
2	1.6467	1.6052	1.5656	1.5278	1.4915
3	2.3216	2.2459	2.1743	2.1065	2.0422
4	2.9137	2.7982	2.6901	2.5887	2.4936
5	3.4331	3.2743	3.1272	2.9906	2.8636
6	3.8887	3.6847	3.4976	3.3255	3.1669
7	4.2883	4.0386	3.8115	3.6046	3.4155
8	4.6389	4.3436	4.0776	3.8372	3.6193
9	4.9464	4.6065	4.303	4.031	3.7863
10	5.2161	4.8332	4.4941	4.1925	3.9232
11	5.4527	5.0286	4.656	4.3271	4.0354
12	5.6603	5.1971	4.7932	4.4392	4.1274
13	5.8424	5.3423	4.9095	4.5327	4.2028
14	6.0021	5.4675	5.0081	4.6106	4.2646
15	6.1422	5.5755	5.0916	4.6755	4.3152
16	6.2651	5.6685	5.1624	4.7296	4.3567
17	6.3729	5.7487	5.2223	4.7746	4.3908
18	6.4674	5.8178	5.2732	4.8122	4.4187
19	6.5504	5.8775	5.3162	4.8435	4.4415
20	6.6231	5.9288	5.3527	4.8696	4.4603
25	6.8729	6.0971	5.4669	4.9476	4.5139
30	7.0027	6.1772	5.5168	4.9789	4.5338
40	7.105	6.2335	5.5482	4.9966	4.5439
50	7.1327	6.2463	5.5541	4.9995	4.5452
75	7.1425	6.2499	5.5555	5.	4.5455
100	7.1428	6.25	5.5556	5.	4.5455

APPENDIX B

U.S. EPA ESTIMATION FORMULAS

REPORTING COST[a]

Description	No./Year	Material $/Occurrence	Labor Hours/ Occurrence	Labor $/Hour
RCRA				
Generators Biennial Report	0.5	5	8	25
LQG Exception Report	0.1–1.5	1	2	25
SQG Exception Report	0–0.1	1	0.25	25
Primary Exporters Exception Report	0.1–1.5	1	2	25
Primary Exporters Annual Report	1	2	2.5	25
TSDF Biennial Report	0.5	5	8–40	25
TSDF Unmanifested Waste Report	0–125	1	1	25
Release, fire, and closure reporting	2	2	5	25
CERCLA: None				
SARA Title III				
Supplemental MSDS Report	0.04–8	4	0.5	20
Requested MSDS Report	Site-specific	1	0.25	20

(Continued)

(Continued)

Description	No./Year	Material $/Occurrence	Labor Hours/ Occurrence	Labor $/Hour
Inventory Report	1	1	5	25
Tier II Reporting by request	0–1	1	5	25
313 Form R reports	1	1	8–40	25
CAA				
Quarterly Compliance & Monitoring Report	4	2	5	25
Performance Test Results Report	4	2	2	25
Opacity Test Results Report	4	2	2	25
Hazardous Pollutant Emission Report	1	2	8	25
Hazardous Pollutant Monitoring System Report	2	1	5	25
CWA				
NPDES Permit Reporting	12	1	1	25
Industrial users continued compliance Report	2	2	5	25
Toxic Standards Annual Compliance Report	1–6	1	5	25
OSHA				
Injury and illness reporting each occurrence	0.05–5	1	1.5	20
Injury and illness annual summary	1	0.25	1	20
Fatality and Hospital Report	0.005–0.5	0	1–10	20
Occupational Injuries and Illness Survey	1–2	0	0.5–3	20

Source: U.S. EPA, *P2 Benefits Manual.* Washington, DC: EPA, 1989.

[a]Reporting cost = (# reports/yr)(labor/report × $/hr + materials $/report).

RECORDKEEPING COST[a]

Description	No./Year	Material $/Occurrence	Labor Hours/ Occurrence	Labor $/Hour
RCRA				
Reports test results, waste analyses records	105–100	1	0.25	9
Exporters reports & notification records	5	1	0.25	9
Manifesting records	0–200	1	0.25	9
Operating record	250	1	0.25	9
CERCLA- none				
SARA Title III				
313 For R reports & calculations	0–2	1	1	9
Notification determination records	0–2	1	1	9
CAA				
Startup, shutdown, & malfunction records	10	1	1	9
Performance test data records	4	1	0.25	9
Opacity test data records	4	1	0.25	9
Hazardous pollutant emission test results	4	1	1	9
Hazardous pollutant monitoring system records	4	1	1	9
CWA				
NPDES monitoring records				
Industrial users/POTW records				
OSHA				
Injury and illness log and summary Medical surveillance program records	1–5	3	0.25	9

Source: U.S. EPA, *P2 Benefits Manual.* Washington, DC: EPA, 1989.

[a]Recordkeeping cost = #/yr [((labor hours/occurrence × labor $/hr) + material $/occurrence)].

NOMINAL PRICES OF WASTE-MANAGEMENT SERVICES (1985–1990)

Disposal Option	Type of Waste	Price per Gallon (unless otherwise noted)		
		1985	1987	1990
Landfill	55-gall drum	$50–$140/drum	$60–190/drum	$90–120/drum
	Bulk	70–140/ton	100–170/ton	60–180/ton
Stabilization	Bulk	Not broken out	$70–270/ton	$210–280/ton
Incineration	Clean liquids, high BTU	$0.10–1.90	$1.30–4.00	$0.40–2.10
	Liquids, low BTU	1.30–4.20	1.30–3.40	1.20–3.60
	Sludges and solids	2.70–4.80	5.40–8.60	3.90–8.10
	Highly toxic liquids	2.10–8.30	2.40–5.00	1.70–4.70
	PCB Liquids	2.50–3.50	2.40–4.30	2.10–3.10
	PCB Solids	4.50–12.50	3.80–8.20	2.50–14.50
Fuel burning	Liquids	Included in incineration	Included in incineration	$0.10–1.25
	Sludges and solids	Included in incineration	Included in incineration	2.95–4.50
Chemical/biological treatment	Aqueous inorganic liquids	Not studied	$0.30–1.20	$0.40–1.50
	Inorganic sludges and solids		0.40–3.50	2.20–5.30
	Aqueous organic liquids			0.50–1.70
Organic recovery	Aqueous organic liquids	$0.00–0.40	$0.20–1.10	$0.40–1.60
	Nonaqueous organic liquids			2.20–5.30
	Oils			0.50–1.70
	Sludges and solids			
Metals recovery	Aqueous liquids	Not studied	Not studied	$0.50–0.80
	Sludges and solids			
Deepwell injection	Aqueous organics, inorganics, wastewaters, oil wastewaters and other toxic organics	$0.10–1.20	$0.10–0.60	$0.20–0.30
Transportation		$0.18–0.22/ton mile	$0.23/ton mile	$0.20–0.30/ton mile
		2.70–4.50/loaded mile	3.30–3.37/loaded mile	2.80–4.60/loaded mile

Source: U.S. EPA, *1990 Survey of Hazardous Waste Management Industries.* Washington, DC: EPA, pp. 2–5.

SOIL AND WASTE REMOVAL AND TREATMENT

$$\text{FL1} = 8.9 \times a \times b \times Q$$

where FL1 is in thousands of dollars.

FL1 gives the general equation for estimating the costs of soil and waste removal and treatment. This same equation applies to potential releases from tanks and transportation (this cost is not applicable to landfill disposal). However, you will need to select different values of parameters a, b, and Q (as provided in what follows) depending on the waste management practice analyzed. For example, if you are analyzing the costs of soil and waste removal and treatment associated with storage in underground tanks, you will set a equal to 2, select a value of b between 0.0001 and 0.1, and set Q equal to the total annual quantity of waste stored in tanks.

Tanks

a = correction factor

= 2 (if underground)

= 1 (if above-ground)

b = fraction of the total annual quantity (treated or stored in tanks) expected to be released

= 0.0001 to 0.1 depending on the type of tank (above-ground or underground; carbon, concrete, stainless, or fiberglass)

Q = total quantity of waste treated or stored in tank (kgal/yr)

Transportation

$a = 1$

b = fraction of the annual quantity (transported) expected to be released

$= 9.5 \times 10^{-8} \times D + 7.6 \times 10^{-6}$ (tanker truck for bulk liquids)

$= 2.4 \times 10^{-6} \times D + 2.9 \times 10^{-4}$ (flatbed truck for drums)

D = distance to treatment or disposal facility (miles)

Q = total quantity of waste transported (kgal/yr)

Landfill Disposal

Not applicable: Because landfills contain a large volume of waste and excavation costs are very high, excavation is unlikely to be the preferred option for remediating a leaking landfill. Leaking landfills result in costs for pumping and treating groundwater.

Source: U.S. EPA, *P2 Benefits Manual.* Washington, DC: EPA, 1989.

GROUNDWATER REMOVAL AND TREATMENT

$$\text{FL2} = a + (b \times c)$$

where FL2 is in thousands of dollars.

FL2 gives the general equation for estimating the costs of groundwater removal and treatment. This same equation applies to potential releases from tanks and landfill disposal (this cost does not apply to transportation practices because releases during transportation generally do not result in groundwater contamination). The values of a and c are given in what follows as a function of other parameters whose values depend on the waste management practice considered.

$$a = \text{capital costs}$$
$$= 91 + [(0.25 \times D \times W) + 0.08D^2 + 137W + 91D + (0.015 \times CC \times V \times W) + (0.005 \times CC \times V \times \text{D})]/1000$$

$$b = \text{O\&M cost}$$
$$= 92 + [11W + 8D + (0.015 \times OM \times V \times W) + (0.005 \times OM \times V \times D)]/1000$$

D = distance to nearest drinking water well (meters)
= 150 to 3200 m

W = width of groundwater plume at facility boundary (meters)

V = groundwater velocity (m/yr)
= 30 to 3000 m/yr

CC = unit capital cost of ground-water treatment (\$/cubic meter/day)

OM = unit operating and maintenance cost of groundwater treatment (\$/yr)/(m^3/day)

Tanks

c = multiplicative factor to determine present value of all O&M costs
= 4 to 8
CC = 440 $/m^3/day
OM = 120 ($/yr)/(m^3/day)
W = 3 to 100 m

Transportation

Not applicable.

Landfill Disposal

$$c = 5 \text{ to } 25$$

$$CC = 340 \text{ \$/m}^3\text{/day}$$

$$OM = 85 \text{ (\$/year)/(m}^3\text{/day)}$$

$$W = 500 \text{ to } 600 \text{ m}$$

Source: U.S. EPA, *P2 Benefits Manual.* Washington, DC: EPA, 1989.

SURFACE SEALING

$$\text{FL3} = CS \times A$$

where FL3 is in thousands of dollars.

FL3 gives the general equation for estimating the costs of surface sealing. This cost applies to landfill disposal and not to tanks and transportation. CS and A are defined in what follows.

Tanks and Transportation

Not applicable.

Landfill Disposal

CS = unit cost of surface sealing landfills (k$/acre)
= 7 to 46 k$/acre

A = area of the landfill (acres)
= 65 to 150 acres

Source: U.S. EPA, *P2 Benefits Manual.* Washington, DC: EPA, 1989.

PERSONAL INJURY

$$\text{FL4} = a \times b$$

where FL4 is in thousands of dollars.

FL4 gives the general equation for estimating the costs of personal injury. This same equation applies to potential releases from tanks and landfill disposal (this cost does not apply to transportation practices). Potential values of a and b follow.

Tanks or Landfill Disposal

a = average claim per person for lost time due to disability and mortality, medical costs related to illness, and medical monitoring costs
= $56,000/person (default value)

b = potentially affected population
= 15,000 (worst-case default value)
= 3,500 (typical-case default value)
= 10 (best-case default value)

Transportation

Not applicable.

Source: U.S. EPA, *P2 Benefits Manual.* Washington, DC: EPA, 1989.

ECONOMIC LOSS

$$\text{FL5} = \text{cost to replace a water supply source}$$

where FL5 is in thousands of dollars.

Tanks and Landfill Disposal

As a first approximation, FL5 may be assumed to be a strict function of the size of the population served. In reality, the cost to replace a water supply source varies with many other parameters such as distance to nearest alternative source of water, cost of purchasing water from an alternative source, and so forth.

Population Served	***Cost to Replace Water Supply Source ($000, 1986)***
50– 99	343
100– 499	412
500– 999	617
1,000– 2,499	1,040
2,500– 4,999	1,812
5,000– 9,999	2,424
10,000– 99,999	10,443
100,000–999,999	51,747
1,000,000+	333,947

Transportation

Not applicable.

Source: U.S. EPA, *Technologies and Costs for the Removal of Fluoride from Potable Water Supplies,* Final Draft, July 1983, pp. 58–78.

REAL PROPERTY DAMAGE

$$FL6 = a \times b \times c$$

where FL6 is in thousands of dollars.

FL6 gives the general equation for estimating the costs of property damage due to contamination of groundwater underlying private property. This same equation applies to potential releases from tanks and landfill disposal (this cost does not apply to transportation practices). The values of a, b, and c are given in what follows as a function of the width of the groundwater plume, W, which is a function of the management practice considered.

a = devaluation factor
= 0.15 to 0.30

b = land value (k$/acre)
= 0.9 to 3.5

c = area of the off-site plume
$= [0.33D^2 + (D \times W) - 0.5W^2]/4047$

D = distance to nearest drinking water well (meters)
= 150 to 3200 m

W = width of groundwater plume at facility boundary (meters)

Tanks

$$W = 3 \text{ to } 100 \text{ m}$$

Transportation

Not applicable.

Landfill Disposal

$$W = 500 \text{ to } 700 \text{ m}$$

Source: U.S. EPA, *P2 Benefits Manual.* Washington, DC: EPA, 1989.

NATURAL RESOURCE DAMAGE

$$\text{FL7} = a \times b$$

where FL7 is in thousands of dollars.

FL7 gives the general equation for estimating the costs of natural resource damage due to contamination of surface water. This same equation applies to potential releases from tanks, transportation, and landfill disposal. The values of a and b are given in what follows as a function of the management practice considered.

a = unit cost of dredging and disposing of contaminated material plus the cost of fish killed (k$/acre)

b = area of surface water contaminated (acres)

Tanks and Landfill Disposal

$$a = 692 \text{ (default value, \$000, 1986)}$$

$$b = 1 \text{ to } 3 \text{ acres}$$

Transportation

$a = 692$ (default value, \$000, 1986)

$b = 1 \times d$ to $3 \times$ d (default values)

$d =$ quantity expected to be released, expressed as a fraction of the annual quantity transported
$= 0.5 \times 10^{-8} \times D + 7.6 \times 10^{-6}$ (tanker truck for bulk liquids)
$= 2.4 \times 10^{-6} \times D + 2.9 \times 10^{-4}$ (flatbed truck for drums)

$D =$ distance to treatment or disposal facility (miles)

Source: U.S. EPA, *P2 Benefits Manual.* Washington, DC: EPA, 1989.

EXPECTED YEAR IN WHICH LIABILITIES ARE INCURRED

$$\text{Year} = \text{time} + (D \times RF)/\text{V}$$

time = expected time lapse between the start of the project and the initial release (years) (ranges from 1 to 20 years)

$D =$ distance to the nearest drinking water well (meters)
$= 150$ (worst-case default value)
$= 1{,}500$ (typical-case default value)
$= 3{,}200$ (best-case default value)

$RF =$ retardation factor for waste constituents (range from 1 to 1,000)

$V =$ groundwater velocity (m/yr)
$= 3{,}000$ (worst-case default value)
$= 300$ (typical-case default value)
$= 30$ (best-case default value)

Tanks

$$\text{time} = 1 \text{ to } 20 \text{ yrs}$$

Transportation

$$\text{time} = 1$$
$$D = 0$$

Landfill Disposal

$$\text{time} = 1 \text{ to } 40 \text{ yrs}$$

Source: U.S. EPA, *P2 Benefits Manual.* Washington, DC: EPA, 1989.

APPENDIX C

CONVENTIONAL AND HIDDEN COSTS

Cost Category	Data Elements	Where Found	Who To Ask
Chemical	Usage rates	Production records	Foreman
	Unit costs	Purchase orders	Billing department
Ancillary	Usage rates	Product specifications	Product engineers
	Unit costs	Purchase orders	Billing department
Storage space	Total square footage	Actual measurement	Maintenance or engineering
	Cost/ft^2	Rental contract	Billing department
Waste treatment	Flow rates	Waste water treatment log sheets	Waste water treatment operator
	Total chemical costs	Purchase orders	Purchasing department
Testing	No. of tests/year	Environmental files	Environmental manager
	Cost per test	Invoices	Accounts payable
Disposal	Type and quantity disposed	Manifests	Environmental manager
	Unit costs of each	Invoices	Accounts payable
Training	No. of people	Training records	Environmental manager
	No. of trainings	Training records	Or Contractor
	Length of training	Training records	Personnel department
	Hourly labor rates	Wage rate sheet	

(Continued)

(Continued)

Cost Category	Data Elements	Where Found	Who To Ask
Personal protective equipment (PPE)	Type and quantity used Cost per item	Stock room or inventory	Environmental manager Purchasing department
Insurance	Type and coverage Premium	Capital budgets Invoices	CFO, accountant Accounts payable
Production	Machine downtime Machine rates Labor rates	Production records Operating budget personnel Records	Production manager Finance department Personnel department
Taxes/fees	Sewer use tax Chemical use tax Water use tax Volume or weight of each taxes item	Water bills Environmental records Water bills Water, chemical Usage records	Accounts payable Environmental manager Local public owned treatment works Production manager Purchasing
Regulatory compliance	Hours of labor for all compliance tasks	Appendix B or environmental management records	Environmental manager
Maintenance labor	Hours of labor Tasks performed	Maintenance log	Maintenance department Shop foreman
Maintenance materials	Amount of materials Costs of materials	Maintenance log Purchase orders	Maintenance department Purchasing department
Water usage	Annual usage rate Costs/gal or ft^3	Flow meters or logs Town water bills	Production manager Accounts payable
Electricity usage	Annual usage rate Cost/kWh	Equipment specifications Utility bills	Production manager Accounts payable
Steam usage	Cost of production Fraction of total used by process	Fuel bills, boiler maintenance logs, count of processes using steam	Accounts payable Maintenance department, plant walk-through

GLOSSARY

Activity Based Costing (ABC): A method of allocating costs based on the types of activities involved with making a product, rather than allocating based on department or broad functional categories such as direct labor and materials. Resources are assigned to activities and activities are assigned to cost drivers based on their use, recognizing a causal relationship between cost drivers and activities.

Amortization: The process of paying off a liability on an installment basis.

Analytical Hierarchy Process (AHP): A process for qualitatively analyzing situations involving many complex and interrelated facts and preferences.

Annual Cash Flow: The sum of cash inflows and outflows for a given period.

Assets: The valuable resources, properties, and property rights owned by an individual or business.

Avoidable Liabilities: Those liabilities that have not been created yet by the firm and can be eliminated or reduced if process or product changes are made. Substituting chemicals sometimes avoids the liabilities of worker exposure to toxic chemicals.

Best Available Control Technology (BACT): Standards of waste treatment technology set by the U.S. EPA to ensure all pollution sources are treated with a technology that provides acceptable reduction of the hazardous components.

Biological Oxygen Demand (BOD): Term used to describe the oxygen consumption potential of an organic compound in water.

Break-Even Point: The point at which the project has paid for itself or its cumulative incremental annual cash flows equal zero.

Capital Budgeting: Also known as investment analysis and financial evaluation, is the process of determining a company's planned capital investments.

Carcinogen: A cancer-causing agent.

Carrying Capacity: The maximum population size that a given geographic area can support without reducing its ability to support the same species in the future.

Cash Flow (of an investment): The dollars coming to the firm (cash inflow) or paid out by the firm (cash outflow) as a result of an investment.

Ceiling Limit: The maximum time-weighted average for exposure to a chemical that cannot be exceeded, even momentarily.

CERES Principles: Formerly known as the Valdez principles and developed by the Coalition of Environmentally Responsible Economies these specify a code of conduct and ethics towards environmental management issues. Formulated after the Exxon *Valdez* oil spill to help prevent similar catastrophes.

Chemical Oxygen Demand (COD): Term used to define the oxygen consumed by an organic chemical as it mixes with water.

Chemical Transportation Emergency Center (CHEMTREC): A service that provides emergency information about how to respond to releases of hazardous chemicals (1-800-424-9300).

Chlorofluorocarbon (CFC): A known ozone-depleting substance.

Chronic Effect: A slowly developing adverse effect, for example, cancerous tumors.

Chronic Toxicity: The capacity of a substance to cause long-term poisonous health effects.

Closure: Deactivating hazardous-waste management portions of a facility.

Code of Federal Regulations (CFR): (1) Are the promulgated regulations as developed by U.S. government. Examples include 40 CFR—EPA Regulations for Hazardous Wastes, Reportable Quantities, SARA, Underground Tanks, etc., often used as models for state environmental regulations;

(2) 49 CFR—DOT Regulations for Hazardous Materials, Marking, Labeling, Placarding and Packaging, etc.

Comprehensive Environmental Response, Compensation, and Liability Act (CERCLA): Also known as the Superfund, which first established legal liability for the generation of wastes at contaminated sites.

Conditionally Exempt Small Quantity Generator (CESQG) of Hazardous Waste

Constituent Concentration in Waste (CCW): The level, usually parts per million of a hazardous substance, the constituent, present in a waste stream.

Contingent Costs: Environmental costs that may be incurred at a late date and are contingent on some event occurring, such as a spill of a toxic chemical.

Cost Driver: Any factor that causes change in the cost of an activity (used within the context of ABC). Activities can have multiple cost drivers.

Created Liabilities: These are the liabilities that result from all the past actions of the facility, such as the disposal of all previously generated hazardous wastes.

Department of Defense (DOD): The agency that regulates the armed services (Army, Navy, Airforce, Marines, etc.).

Department of Energy (DOE): The agency that regulates atomic energy research and national scientific laboratories such as Los Alomos and Sandia.

Department of Transportation (DOT): The agency that regulates transportation of hazardous materials, including hazardous wastes over roads, water, or by air.

Discount Rate: The interest rate at which money can be invested or borrowed. In a TCA, the discount rate is used in the net present value calculations to determine the value of future expenditures in the present year. The discount rate is expressed as a percentage.

Discounted Cash Flow (DCF) Analyses: A number of financial decision-making tools that account for the time value of money by discounting or reducing the value of future cash payments.

Ecological Footprint: The land and water area that would be required to support a defined human population and material standard indefinitely.

Emergency Planning and Community Right-to-Know Act of 1986 (EPCRA): Also known as SARA Title III.

Environmental Accounting: This term has three distinct meanings: (1) within the context of national income accounting, it refers to the accounting of natural resources and their consumption, extent, and quality; (2) within the context of financial accounting, it refers to documenting environmental compliance costs using GAAP for external audiences; (3) within the context of management accounting, it serves as a decision-making tool and performance indicator and includes the processes of TCA.

Environment, Health & Safety (E, H & S): Term generally applied to the position of a manager responsible for oversight of worker health and safety and management of pollution from an industrial facility.

Environmental Liabilities: An umbrella term used to refer to different types of environmental costs, including the costs for remediating contaminated soils, property and personal injury claims, and loss of economic use of lands.

Environmental Protection Agency (EPA): The agency that regulates industrial pollution entering the nation's waters, air or land. Also supports a variety of voluntary, pollution prevention initiatives such as Green Lights (energy conservation) and the 33/50 program (reduction in use of specify chemicals).

Equity: The monetary value of a property or business that exceeds the claims and/or liens against it by others.

Extremely Hazardous Substance (E'S): All substances listed in 40 CFR Part 355, Appendixes A and B.

Federal Insecticide, Fungicide and Rodenticide Act (FIFRA)

Federal Register: A daily U.S. government publication that contains final and proposed changes to regulations of the U.S. government agencies, including all EPA environmental regulations.

Form R: Toxic Chemical Release Inventory Reporting form required by SARA Title III, Section 313, which tracks annual releases of listed chemicals if the facility uses them over a regulatory threshold limit.

General Permit: A National Emissions Standards, for Hazardous Air Pollutants permit authorizing a category of water discharges under the Clean Water Act within a geographical region.

Generally Accepted Accounting Principles (GAAP): A set of accounting practices, tenets, and principles that come from the Financial Accounting

Standard Board, Accounting Principles Board, and the American Institute of Certified Public Accountants, and commonly used by accountants in financial reporting.

Hazardous Air Pollutant (HAP): A substance released to the air outside a manufacturing facility which has been determined to cause adverse health effects.

Hazardous Chemical: A material that exhibits any physical or any health hazard. Hazardous chemicals are regulated under 29 CFR 1910.1200 (OSHA)—Hazard Communication and 40 CFR Part 370—Hazardous Chemical Reporting.

Hazardous Material: Any material listed in 40 CRF 172.101 (EPA). All EPA hazardous wastes are hazardous materials.

Hazardous Substance: Substances listed in the Appendix to 49 CFR 172.101 (DOT) Hazardous-Materials Table when in a quantity in one container that equals or exceeds the Reportable Quantity (RQ), and substances listed in 40 CRF 302.4.

Hazardous Waste: Any material meeting the definition of New Jersey's hazardous-waste regulations. They include listed (F, K, U, P, X, C numbers) and characteristics (D numbers) hazardous wastes.

Hurdle Rate: An internally defined interest rate that all the internal rate of return of any capital investment is expected to exceed.

Incremental Cash Flow: The difference between cash flows from one option and another. In TCA, it is the difference between the annual operating cash flows of the current process and those of the proposed process.

Insured: The person or business covered by the insurance policy.

Internal Rate of Return (IRR): The discount rate at which the net present value of the investment is equal to zero. Alternatively, it is the rate at which the present value is equal to the capital costs of the project.

ISO 14000: Standards presently being developed for international use in implementing a quality environmental management system based on the principles of ISO 9000, which provides third-party certification of uniform production quality for manufacturers.

Joint and Several Liability: A liability is joint and several when a person may sue one or more of the parties to the liability separately, or has the option to sue all of them together. Under CERCLA, each potentially responsible party (PRP) is jointly and severally liable for all costs associated with the site clean-up if such costs cannot be allocated among the PRPs.

Land Disposal: Placement/disposal of wastes in a landfill, surface impoundment, waste pile, injection well, land treatment facility, salt dome or salt bed formation, underground mine or cave, or concrete vault or bunker for disposal purposes.

Land Disposal Restrictions (LDR): Regulations banning the landfill disposal of certain listed chemicals and wastes or specifying the preferred method of their disposal (i.e., incineration or verification).

Large Quantity Generator (LQG) of Hazardous Waste: A facility that generates more than 11,000 kg (4.75 fifty-five-gall drums) per month of hazardous wastes.

LC_{50}: Lethal concentration of a chemical that when applied to a group of living organisms will kill 50% of them.

Life Cycle Assessment: A holistic approach to identifying the environmental consequences of a product, process, or activity. By itself, life cycle assessment focuses on environmental impacts, not costs. The assessment includes all aspects of design, materials selection and use, manufacture, marketing, transportation, use of the product, and end disposal.

Local Emergency Planning Committee under SARA Title III (LEPC): A committee of local officials (fire department, townhall, hospital) and industry representatives responsible for developing emergency response and evaluation plans for the communities surrounding the local industries. This committee uses information provided by the industry, such as type and quantity of chemicals present at the facility, to determine the safe evacuation radii in the event of a spill, explosion or other emergency.

Manifest: Shipping papter used in documenting all shipments of hazardous waste.

Material Safety Data Sheet (MSDS): A chemical information sheet required by the Hazard Communications Standard (OSHA). The data provided by the chemicals manufacturer is used to respond to emergency conditions such as the spill of a chemical, and ingestion or contact with the chemical. It also alerts the chemical user to potential long-term toxicity effects, storage requirement and the chemicals reactivity with other materials.

Materials Balance: A process of analyzing the chemical and material inputs and outputs of a manufacturing process or facility. Usually, a process flow diagram is included to graphically show where all the inputs and outputs are and where they go.

Montreal Protocol: An international environmental agreement to control chemicals that deplete the stratospheric ozone layer. The protocol, which was renegotiated in 1990, calls for a phaseout of CFCs, halons, and carbon tetrachloride by the year 2000; a phaseout of methyl chloroform by 2005; and provides financial assistance to help developing countries make the transition from ozone-depleting substances.

Mutagen: An agent that causes biological mutations.

National Emissions Standards for Hazardous Air Pollutants (NESHAP): These standards establish levels of hazardous air pollutants that can exist in the outdoor air with air harm to the public.

National Pollutant Discharge Elimination System (NPDES): NPDES permits must be obtains by companies that discharge wastewater to surface water bodies (rivers, streams, ponds, lakes, estuaries).

National Priorities List (NPL): A list compiled by the EPA of the topmost contaminated sites in America. These sites are prioritized for litigation and remediation by the federal government and potentially responsible parties.

National Response Center (NRC): A branch of the U.S. Coast Guard. Spill of hazardous substances in excess of the reportable quantities listed in 40 CFR part 302 must be reported to the National Response Center.

Net Present Value (NPV): A calculation performed to adjust the value of an investment so that all future payments can be compared in current dollar amounts. The future payments are "discounted" by the discount rate to adjust for risk, inflation, debt, and equity. This is one financial decision tool from the family of discounted cash flow analyses.

Occupational Safety and Health Administration (OSHA): This agency regulates working conditions within an industrial facility its regulations cover noise, vibration safety markings, labels of chemicals and the worker's right to know about the health effects of the chemicals they are exposed to.

Overshoot: Growth beyond an area's carrying capacity, leading to crash, an example would be any city that has excessive traffic congestion, air pollution imports of food supply and drinking water purity issues.

Payback Period: The amount of time required for an investment to pay back the initial capital costs. This is not a discounted cash flow analysis because it does not adjust for the time value of money.

Personal Protective Equipment (PPE): This is equipment used by workers to protect themselves from exposure to hazardous materials. PPE would include gloves, masks, respirators, boots, aprons, etc.

PM-10: A standard for measuring the amount of solid or liquid matter suspended in the atmosphere ("particulate matter"). Refers to the amount of particulates less than 10 micrometers in diameter.

Pollution Legal Liability (PLL): Insurance coverage for third-party legal liability arising from pollution conditions that emanate from covered locations, including defense costs and remediation expenses for on-site cleanup.

Pollution Prevention: The prevention of waste and pollution releases at the source through the use of input substitution, process modification, process modernization, preventative maintenance, worker training, or integral recycling.

Potentially Responsible Party (PRP): A company that previously sent hazardous wastes to a landfill or disposal site that has now become a Superfund hazardous-waste site.

Publicly Owned Treatment Works (POTW): A municipal wastewater treatment facility.

Remedial Investigation and Feasibility Study (RI/FS): This is the first step in the process of remediating a site contaminated with hazardous waste. It involves exploratory sampling of soils and water to determine the scope and magnitude of contamination and potential obstacles to clearing up the side.

Reportable Quantity (RQ)

Resource Conservation and Recovery Act: Federal statute that granted the EPA the authority to regulate hazardous waste.

Securities and Exchange Commission (SEC): This agency sets policies on investing, trading and reporting of financial information for publicly held companies.

Small Quantity Generator (SQG) of Hazardous Waste: One that creates less than 100 kg of hazardous waste per month.

Societal Costs: The costs of a company's impacts on the environment and society for which a company is not financially responsible. These costs are also referred to as external costs.

Source Reduction: Elimination of wastes at the source of generation, usually the production process and before the wastes leave the facility.

Spill Prevention Control and Countermeasure (SPCC): This phrase refers to a plan that provides instructions on how to respond to an oil spill, who should be contacted and who should clean up the spill. Each industrial facility storing oil, on-site in excess of the regulatory thresholds must prepare such a plan and submit it to the U.S. EPA.

Strict Liability: Strict liability, which can only be authorized by statute, imposes liability on a person simply as a result of whether that person does or does not perform a certain act. It is irrelevant what the intent was or whether negligence is involved. The mere act itself constitutes liability. In the CERCLA context, the simple act of having arranged for waste disposal of a hazardous substance is enough to create liability for cleanup, even if everything was done in compliance with the law.

Superfund Amendments and Reauthorization Act (SARA): SARA Title III is the Emergency Reporting and Community Right-to-Know portion of this act.

Supplemental Environmental Project (SEP): A project conducted by a company out of compliance with environmental regulations and facing administrative penalties. The costs of this project offset a portion of the penalty, usually not more than 50% of the total penalty amount.

Teratogenic: Causing fetal malformations.

Theory of Constraints (TOC): A management theory that proposes no manufacturing operation can operate faster than the slowest step (constraint) of that process, and that a business has but one goal—to make money.

Threshold Limit Value (TLV): These values are limits for worker exposure to hazardous materials (airbone).

Time-Weighted Average (TWA): The concentration of an airborne chemical that a person can be exposed to for a period of time and suffer no ill effects.

Total Cost Assessment (TCA): A decision-making tool that derives the total cost of conducting business in a particular manner, by including environmental costs traditionally placed into the overhead operating budget of a corporation. TCA usually involves the use of discounted cash flow analyses to compare two or more environmental projects.

Total Quality Management (TQM): A set of activities to promote continuous improvement and exceeding customers' satisfactions.

Toxic Chemical: A chemical listed on 40 CFR Part 372, Subpart D.

Toxicity Characteristic Leachate Procedure (TCLP): The primary method for identification of hazardous waste by simulating conditions within a landfill.

Treatment, Storage, and Disposal Facility (TSDF): Usually, a final destination for hazardous wastes or a facility that has waste disposal sites on its property such as a sludge lagoon or settling pit.

Very Small Quantity Generator (VSQG) of Hazardous Waste: A facility that generates less than 100 kg of hazardous waste in one month.

Weighted Average Cost of Capital (WACC): A discount rate composed of multiple factors affecting the ability of a company to invest in a project.

BIBLIOGRAPHY

Aetna Casualty & Surety Company v. *Georgia Tubing Corporation*, Dockett No. 95-5072, August 19, 1996.

Anderson, Donald, Director of Environmental Liability Management Institute, Toronto, interview by author on March 14, 1997.

Arco Corp., *1995 Annual Report Financials*, Arco, 1995, web site: http://www.arco.com:80/Corporate/numbers/ar95/note1.htm.

Bucchieri, Richard, Environmental Geologist, Haley & Aldrich, Cleveland, interview by author on April 3, 1997.

Canadian Institute of Chartered Accountants, *Environmental Costs and Liabilities: Accounting and Financial Reporting Issues*, Toronto: CICA, 1993.

Capezio, Peter, *Taking the Mystery Out of TQM*, Hawthorne, NJ: Career Press, 1993.

CEEM, *International Environmental Systems Updates*, Farifax, VA: CEEM, June and July, 1995.

Connecticut Department of Environmental Protection, Press Releases from Q River Oil Spill, Hartford: CTDEP, Office of Communications, October 18–November 8, 1995.

CSX Corp., *1995 Annual Report: Note to Consolidated Financials*, CSX, 1995, web site: http://www.csx.com:80/docs/95 annrep/notes 13.html.

Daly, Herman, and John Cobb, *For the Common Good*, Boston: Beacon Press, 1989.

Dempsey, Michael, *The Development of a Theory of Corporate Investment Decision Making*, Leeds, England: School of Business and Economics, University of Leeds, 1993.

Denton, Keith D., *Enviro-Management—How Smart Companies Turn Environmental Costs Into Profits*, Englewood Cliffs, NJ: Prentice Hall, 1994.

Diekmann, James, and Gary Baker, *EPA/Corps of Engineers Joint Guidance Document for Cost Risk Analysis for HTW Remediation Projects*, Washington, DC: U.S. EPA, Office of Solid Waste and Emergency Response, October 1992.

Donnahoe, Alan S., *What Every Manager Should Know About Financial Analysis*. New York: Simon & Schuster, 1989.

DuVall, Barry J., *Contemporary Manufacturing Processes*, South Holland, IL: Goodheart-Wilcox Co., 1996.

Dinoia, Michele, Supervising Sanitary Engineer, Connecticut Department of Environmental Protection, Water Bureau, Hartford, interview by author on March 18, 1997.

ECS Underwriting, *The Environmental Insurance Market-Rapid Emergence to Stable Growth*, Exton, PA: ECS Underwriting, Inc., 1996.

Evans, Gordon, Economist, U.S. EPA, Office of Research and Development, Cincinnati, interview by author on March 18, 1997.

Federal Register, "Interim Revised EPA Supplemental Environmental Projects Policy Issued," Vol. 60, No. 90, May 10, 1995, p. 24860.

Goldberg, Terry, Director of Northeast Waste Management Officials Association, Boston, interview by author on January 23, 1997.

Giuntini, Ron, CPIM, "Helping Your Company Go Green," *Management Accounting*, Volume 75, No. 8, February 1994, p. 43.

Grimm, Robert, Vice President and Chief Information Officer, Insurance Service Office, New York, interview by author on March 26, 1997.

GZA GeoEnvironmental, Inc., *Hazardous Waste Rules and Regulations—A Comprehensive Compliance Seminar and Reference Manual*, Newton, MA: GZA, 1995.

———. *Selected 1994 Massachusetts Toxics Use Reduction Plans*, Vernon, CT: GZA, 1997.

Hatfield, Scott, "Re: Wrap Up on Environmental Liabilities," public memorandum, July 24, 1995.

International Chamber of Commerce, *Business Brief No. 1: Economics and the Environment*, Paris: ICC Task Force, 1992.

Jenner & Block, "Second Circuit Court of Appeals Holds Certain Pre-Petition Environmental Claims Dischargeable in Bankruptcy," *Land, Air and Water News*, Fall 1991, pp. 1–3.

Katzen, Amelia-Welt, Supplemental Environmental Projects Managers, U.S. EPA Region I, Office of Environmental Stewardship, Boston, interview by author on March 27, 1997.

Kennedy, Mitchell L., *The Cost of Changing: A Total Cost Analysis of Solvent Alternatives*, Lowell, MA: Massachusetts Toxics Use Reduction Institute, 1993.

Kieso, Donald E., *Intermediate Accounting*, 5th ed., New York: John Wiley, 1986.

Krueze, Jerry G. "ABC and Life-Cycle Costing for Environmental Expenditures," *Management Accounting*, Volume 75, No. 8, February 1994, p. 38.

Laughlin, Sean, Software Support Technician for Expert Choice, Pittsburgh, interview by author on March 11, 1996.

Libber, Jonathan, U.S. EPA BEN/ABLE Coordinator, U.S. Environmental Protection Agency, Office of Enforcment and Compliance Assurance, Washington, D.C., interviews by author on March 29 and April 3, 1997.

McDonough, William, *The Hanover Principles/Design for Sustainability*, Oakland, CA: William McDonaugh Architects, 1992.

McInerney, Francis, *The Total Quality Corporation*, New York: Penguin Books, 1995.

McKinnon, Sharon M, *The Information Mosaic—How Managers Get the Information They Really Need*, Cambridge, MA: Harvard Business School Press, 1992.

National Center for Manufacturing Sciences, *Activity Based Costing/Activity Based Management: The Next Frontier*, Ann Arbor, MI: NCMS, 1993.

Northeast Waste Management Officials Association *Costing and Financial Analysis of Pollution Prevention Projects: A Training Packet*, Boston, MA: NWMOA, 1994.

Pezzey, John, *Economic Analysis of Sustainable Growth and Sustainable Development*, Environmental Working Paper No. 15, Geneva: World Bank, 1989.

PRC Environmental Management, *Total Quality Management/Pollution Prevention Integration Study for the Printed Circuit Board Manufacturing Industry*, Los Angeles: PRC, 1993.

Reiser, Matthew, Associate Account Manager, Special Liability Coverage Unit, Travelers Insurance, interview by author on March 27, 1997.

Resource Renewal Institute, *What Are Green Plans?* San Francisco: RRI, 1994.

Reynolds, Anne, Environmental Manager for Lucien Technologies, Andover, Massachusetts, interview by author on March 6, 1997.

Ruch, Susan, "Contingent Environmental Liabilities: Trend Toward More Detailed Disclosure in Financial Statements," Internet Web Page, Testa, Hurwitz & Thibeault, LLP, 1994, web site: http://www.tht.com:80/VUasrtSu94cont.htm.

Schorr, Juliet, *The Overworked American*, Boston: HarperCollins, 1991.

Schenck, Rita, "SEC and GAAP Article—Email Response," February 8, 1997, web site: http://enfo.com:80/NAEP/MailLists/Archives/0161.html.

Securities and Exchange Commission, *1979 SEC Lexis 621—Environmental Disclosure*, Washington, DC: SEC, 1979.

———. *Regulation S-K, Item 101 (c) (1) (xii)*, Washington, DC: SEC, 1973.

Smith, Jeffry J., *Theory of Constraints and MRP II: From Theory to Results*, Peoria, IL: Bradley University, 1994.

Sternberg, Robert J., *In Search of the Human Mind*, Orlando, FL: Harcourt Brace, 1995.

Thiesen, G. J. and W. J. Fabrycky, *Engineering Economy*, Englewood Cliffs, NJ: Prentice Hall, 1984.

Thompson, Charles, *"Yes, But . . ."—The Top 40 Killer Phrases and How You Can Fight Them*, New York: Harper Business, 1993.

Turney, P.B.B. "Activity-Based Costing: A Tool for Manufacturing Excellence," *Target*, Summer 1989, pp. 13–19.

———. "What an Activity-Based Cost Model Looks Like," *Journal of Cost Management*, Winter 1992, pp. 54–60.

Vasil, Greg, Director of Environmental Strike Force, Massachusetts Department of Environmental Protection, Boston, interview by author on March 26, 1997.

Von Oech, Roger, *A Kick in the Seat of the Pants*, New York: Harper Perennial, 1986.

Wackernagel, Mathis and William Rees, *Our Ecological Footprint—Reducing Human Impact on the Earth*, B.C.: New Society, 1996.

Walker, Thomas G., Vice President, WRAMSCO, Inc., Donners Grove, Illinois, interview by author on March 18, 1997.

Wall, Francis J. *Statistical Data Analysis Handbook*, New York: McGraw-Hill, 1986.

Williams, Bill, *A Sampler on Sampling*, New York: John Wiley, 1978.

Williams, Georgina, "Cleaning Up Our Act: Accounting for Environmental Liabilities," *Management Accounting*, Volume 75, No. 8, February 1994, p. 30.

Wenick, Neil, Vice President of Marketing, ECS, Inc., Exton, Pennsylvania, interview by author on March 16, 1997.

U.S. EPA, Office of Enforcement and Compliance Monitoring, "Guidance on Calculating After Tax Net Present Value of Alternative Payments," public memorandum, Washington, D.C., October 28, 1986.

———. "Policy on the Use of Supplemental Environmental Projects in EPA Settlements," public memorandum, Washington, D.C., February 12, 1991.

U.S. EPA, Office of Pollution Prevention and Toxics, *Environmental Accounting Case Studies: Green Accounting at AT&T*, Document EPA 742-R-95-003, Washington, DC: EPA, 1995.

———. *Environmental Cost Accounting for Capital Budgeting: A Benchmark Survey of Management Accountants*, Document EPA 742-R-95-005, Washington, DC: EPA, 1995.

U.S. EPA, Office of Policy Planning and Evaluation, *Pollution Prevention Benefits Manual*, Vols. I and II, Document EPA/230/R-89/100, Washington, DC: EPA, 1989.

———. *Incorporating Environmental Costs and Considerations into Decision Making: Review of Available Tools and Software*, Document EPA 742-R-95-006, Washington, DC: EPA, 1996.

———. *An Introduction to Environmental Accounting As a Business Management Tool: Key Concepts and Terms*, Document EPA 742-R-95-005, Washington, DC.: EPA, 1995.

U.S. EPA, Office of Research and Development, *Life Cycle Design Guidance Manual*, Document EPA/600/R-92/226, Washington, DC: EPA, 1993.

———. *A Primer for Financial Analysis of Pollution Prevention Projects*, Washington, DC: EPA, 1993.

Westinghouse, Electric Corp., *Westinghouse Electric Corporation 1995 Annual Report*, WEC, web site: http://www.westinghouse.com:80/corp/95annual/page_050.htm#10.

INDEX